Approximate Methods in Optimization Problems

Modern Analytic
and Computational
Methods *in* Science
and Mathematics

A GROUP OF MONOGRAPHS
AND ADVANCED TEXTBOOKS

Richard Bellman, EDITOR
University of Southern California

Published

1. R. E. Bellman, Harriet H. Kagiwada, R. E. Kalaba, and Marcia C. Prestrud, Invariant Imbedding and Radiative Transfer in Slabs of Finite Thickness, 1963

2. R. E. Bellman, Harriet H. Kagiwada, and Marcia C. Prestrud, Invariant Imbedding and Time-Dependent Transport Processes, 1964

3. R. E. Bellman and R. E. Kalaba, Quasilinearization and Nonlinear Boundary-Value Problems, 1965

4. R. E. Bellman, R. E. Kalaba, and Jo Ann Lockett, Numerical Inversion of the Laplace Transform: Applications to Biology, Economics, Engineering, and Physics, 1966

5. S. G. Mikhlin and K. L. Smolitskiy, Approximate Methods for Solution of Differential and Integral Equations, 1967

6. R. N. Adams and E. D. Denman, Wave Propagation and Turbulent Media, 1966

7. R. L. Stratonovich, Conditional Markov Processes and Their Application to the Theory of Optimal Control, 1968

8. A. G. Ivakhnenko and V. G. Lapa, Cybernetics and Forecasting Techniques, 1967

9. G. A. Chebotarev, Analytical and Numerical Methods of Celestial Mechanics, 1967

10. S. F. Feshchenko, N. I. Shkil', and L. D. Nikolenko, Asymptopic Methods in the Theory of Linear Differential Equations, 1967

11. A. G. Butkovskiy, Distributed Control Systems, 1969

12. R. E. Larson, State Increment Dynamic Programming, 1968

13. J. Kowalik and M. R. Osborne, Methods for Unconstrained Optimization Problems, 1968

14. S. J. Yakowitz, Mathematics of Adaptive Control Processes, 1969

15. S. K. Srinivasan, Stochastic Theory and Cascade Processes, 1969

16. D. U. von Rosenberg, Methods for the Numerical Solution of Partial Differential Equations, 1969

17. R. B. Banerji, Theory of Problem Solving: An Approach to Artificial Intelligence, 1969

18. R. Lattès and J.-L. Lions, The Method of Quasi-Reversibility: Applications to Partial Differential Equations. Translated from the French edition and edited by Richard Bellman, 1969

19. D. G. B. Edelen, Nonlocal Variations and Local Invariance of Fields, 1969

20. J. R. Radbill and G. A. McCue, Quasilinearization and Nonlinear Problems in Fluid and Orbital Mechanics, 1970

21. W. Squire, Integration for Engineers and Scientists, 1970

23. T. Hacker, Flight Stability and Control, 1970

24. D. H. Jacobson and D. Q. Mayne, Differential Dynamic Programming, 1970

25. H. Mine and S. Osaki, Markovian Decision Processes, 1970

27. E. D. Denman, Coupled Modes in Plasmas, Elastic Media, and Parametric Amplifiers, 1970

30. D. H. Moore, Heaviside Operational Calculus: An Elementary Foundation

32. V. F. Demyanov and A. M. Rubinov, Approximate Methods in Optimization Problems

34. C. J. Mode, Multitype Branching Processes: Theory and Applications, 1970

In Preparation

22. T. Parthasarathy and T. E. S. Raghavan, Some Topics in Two-Person Games

26. W. Sierpinski, 250 Problems in Elementary Number Theory

28. F. A. Northover, Applied Diffraction Theory

29. G. A. Phillipson, Identification of Distributed Systems

31. S. M. Roberts and J. S. Shipman, Two-Point Boundary Value Problems: Shooting Methods

33. S. K. Srinivasan and R. Vasudevan, Introduction to Random Differential Equations and Their Applications

35. R. Tomovic and M. Vukobratovic, General Sensitivity Theory

36. J. G. Krzyz, Problem Book in Artificial Functions

37. W. T. Tutte, Introduction to the Theory of Matroids

Approximate Methods

Translated from the Russian by
SCRIPTA TECHNICA, INC.
Translation Editor: George M. Kranc
Department of Electrical Engineering
City University of New York

Vladimir F. Demyanov

and

Aleksandr M. Rubinov

Leningrad State University
Leningrad, U.S.S.R.

n Optimization Problems

American Elsevier
Publishing Company, Inc.
NEW YORK · 1970

ORIGINALLY PUBLISHED AS
Priblizhennyye metody resheniya ekstremal'ykh zadach
Leningrad University Press, 1968
The English language edition of this book
contains numerous changes as well as additions
which have been made by the author
after the Russian edition was published.
The present work may be considered
as a SECOND EDITION.

AMERICAN ELSEVIER PUBLISHING COMPANY, INC.
52 Vanderbilt Avenue, New York, N.Y. 10017

ELSEVIER PUBLISHING COMPANY, LTD.
Barking, Essex, England

ELSEVIER PUBLISHING COMPANY
335 Jan Van Galenstraat, P.O. Box 211
Amsterdam, The Netherlands

International Standard Book Number 0-444-00088-7

Library of Congress Card Number 75-135057

Manufactured in the United States of America

CONTENTS

PREFACE

The present book takes up methods of investigating and solving certain nonlinear extremal problems.

The central position is occupied by Chapter 2, which is devoted to the necessary conditions for an extremum, and Chapter 3, which gives algorithms for testing points suspected of being extrema. Chapter 1 gives a description of the mathematical concepts used in the book. The results of Chapters 2 and 3 are used in Chapter 4 to investigate a number of problems in optimal-control theory. Finally, Chapter 5 is devoted to finite-dimensional problems.

The book is written in the language of functional analysis. Chapter 1 gives a brief summary of certain material from functional analysis that will be used later on. Nonetheless, it is assumed that the reader is familiar with the fundamentals of that science.

In recent years, there have been many interesting approaches to the study of the problems considered in this book (see, for example, [3, 25, 27, 33, 34, 38, 39, 41-44, 46, 47, 49, 56-58]). However, the small size of the book compelled us to confine ourselves to an exposition of those questions that we ourselves had been studying. Therefore, the book does not in any way pretend to treat fully all the topics that interest us. We were also compelled to give certain results in Chapters 4 and 5 without proof.

We wish to express our deep gratitude to Yuri Fedorovich Kazarinov, who undertook the difficult obligations of editor of the book, and also to Vladimir Ivanovich Zubov and to Gleb Pavlovich Akilov, without whose attention and criticisms this book would very likely not have been published. The abstracts of our lectures, which were written by T. K. Vinogradova, considerably facilitated the work on the book.

Vladimir F. Demyanov

Aleksandr M. Rubinov

CHAPTER 1

Introduction

1. Certain results from functional analysis

Here, we shall give a brief exposition, for the most part without proofs, of those results from functional analysis that will be essential in what follows. However, it is assumed that the reader is familiar with the fundamentals of functional analysis (see, for example, [32, 40]).

1^0. In this book, we shall consider only real normed spaces. Particularly, we shall make use of the spaces $C(E)$ and $L^p(E)$, where $1 \leqslant p \leqslant \infty$.

Let E denote a close bounded subset of a finite-dimensional space. We denote by $C(E)$ the space of all real functions that are continuous on E, with norm given by

$$\|x\| = \max_{t \in E} |x(t)|.$$

Now, let E denote a Lebesgue-measurable set of either finite or infinite measure. The space $L^p(E)$, where $(1 \leqslant p < \infty)$, consist of all functions that are p-summable on E. Functions that coincide except on a set of measure zero are considered as a single element of this space. The *norm* is defined by

$$\|x\| = \left(\int_E |x(t)|^p \, dt \right)^{1/p}.$$

The space $L^\infty(E)$ consists of all measurable almost-everywhere-bounded functions defined on E. (As before, functions that differ only on a set of measure zero are considered as a single element of $L^\infty(E)$.) If $x \in L^\infty(E)$, then

$$\|x\| = \operatorname{vrai}\max_{t\in E} |x(t)|.$$

Let X denote a normed space. Following the usual custom, we denote by X^* the *dual* of X, that is, the space whose elements are linear[1] functionals defined on X with norm defined by

$$\|f\| = \sup_{\|x\|\leqslant 1} |f(x)|.$$

If $1 \leqslant p < \infty$, then the space $(L^p(E))^*$ is isometric to the space $L^q(E)$, where p and q are related by $1/p + 1/q = 1$ (if $p=1$, then $q=\infty$). This isometry is achieved as follows: corresponding to every $f \in (L^p(E))^*$ there exists exactly one element $y \in L^q(E)$ such that, for every $x \in L^p(E)$,

$$f(x) = \int_E x(t)\, y(t)\, dt.^2 \tag{1.1}$$

Conversely, every element $y \in L^q(E)$ generates in accordance with formula (1.1) a functional $f \in (L^p(E))^*$ (Theorem 1 (2.VI) in [32]). Here,

$$\|y\|_{L^q(E)} = \|f\|_{(L^p(E))^*}.$$

In what follows, we shall frequently indentify the functional f and its generative element y; we shall also often identify the spaces $(L^p(E))^*$ and $L^q(E)$.

[1]A functional f is said to be *linear* if it is additive (that is, $[f(x+y) = f(x) + f(y)]$ and continuous.

[2]The formulas in this book are numbered with the aid of a double index: the first part of it indicates the number of the paragraph, and the second indicates the number of the formula in the paragraph. When references are made to formulas in a different chapter, the number of the chapter is also indicated.

Let us describe the space dual to $C(E)$, where E is a closed bounded set. We denote by Σ the set of all measurable subsets of E. We shall say that a countably additive function (or measure) μ is defined on Σ if to every $e \in \Sigma$ there corresponds a real number $\mu(e)$ and if for every disjoint sequence of sets e_k in Σ we have

$$\mu\left(\bigcup_{k=1}^{\infty} e_k\right) = \sum_{k=1}^{\infty} \mu(e_k).$$

A measure μ is said to be *finite* on E if, for every e in Σ, we have $-\infty < \mu(e) < \infty$. The finiteness of μ on E implies that, for every $e \in \Sigma$,

$$\sup \sum_{k=1}^{n} |\mu(e_k)| < \infty. \tag{1.2}$$

Here, the supremum is taken over all possible partitions of e into a finite number of subsets:

$$e = \bigcup_{k=1}^{n} e_k;\ e_i \cap e_j = \varnothing\ (i \neq j).$$

The quantity (1.2) is called the ***total variation*** of μ on e and it is denoted by var $\mu(e)$. We denote by $V(E)$ the normed space of all measures that are finite on E with the norm defined by $\|\mu\| = \text{var}\,\mu(E)$.

It can be shown that $V(E)$ is isometric to $(C(E))^*$. Also, a functional $f \in (C(E))^*$ and the measure $\mu \in V(E)$ corresponding to it are related by

$$f(x) = \int_E x\,d\mu$$

where $x \in C(E)$. In what follows, we shall identify f and μ and also $V(E)$ and $(C(E))^*$.

2°. Let X denote a normed space and let x and y denote members of X. The *line segment* connecting all points x and y is the set of points such that $\{y+\alpha(x-y) \mid \alpha \in (0, 1)\}$. The *ray* with origin at the point y in the direction $x-y$ is the set of points such that $\{y+\alpha(x-y) \mid \alpha > 0\}$. (In the case of a ray with origin at 0, we shall usually omit mention of this origin.) The line passing through x and y is the set of points such that $\{y+\alpha(x-y) \mid -\infty < \alpha < \infty\}$. We shall say that a set Ω is *convex* if, for any two points x and y in it, it contains the segment connecting those points. (In other words, if x and y are members of Ω, then the point $\alpha x+(1-\alpha)y$ also belongs to Ω for every $\alpha \in [0, 1]$.) A convex set Ω is said to be *corporeal* if it has at least one interior point. We shall denote by $\mathring{\Omega}$ the set of all interior points of Ω (this set being called the *interior* of Ω). We shall say that a convex corporeal set is *strictly* convex if

$$\alpha x+(1-\alpha) y \in \mathring{\Omega}.$$

for every x and y in Ω and every α in (0, 1). If Ω is a corporeal convex set, then a point x in Ω that is not an interior point of Ω is called a *boundary point* of Ω. The set Ω' of all boundary points of Ω is called the boundary of Ω. Any linear subset will serve as an example of a convex subset of X. As one can easily show, the sphere $S_r(x)$ of radius r with its center at the point x, that is, the set

$$x\,(S_r(x)=\{y \in X \mid \|x-y\| \leqslant r\})$$

is a convex corporeal set. Here,

$$(\mathring{S}_r(x))=\{y \in S_r(x) \mid \|x-y\| < r\};$$
$$(S_r(x))'=\{y \in S_r(x) \mid \|x-y\| = r\}.$$

The same sphere in the space $L^p(E)$, where $1 < p < \infty$, is a strictly convex set.

Suppose that $\Omega \subset X$. We shall call the set

$$\left\{y \in X \mid y=\sum_{k=1}^{n} \alpha_k x_k,\ \ \alpha_k \geqslant 0,\ \ x_k \in \Omega,\ k=1, \ldots, n;\right.$$
$$\left.\sum_{k=1}^{n} \alpha_k=1,\ n=1, 2, \ldots\right\},$$

the *convex hull* of the set Ω. One can show that the convex hull of Ω

is contained in an arbitrary convex set containing Ω and, in this sense, it is the smallest convex set containing Ω.

We shall refer to a set K as a cone (or cone with vertex at 0) if, for every point x in K, the ray in the direction x is contained in K (in other words, if the relation $x \in K$ implies $\alpha x \in K$ for every $\alpha > 0$). We shall refer to a set $K(y)$ as a cone with its vertex at y if, for every point x in it, the set $K(y)$ contains the ray with origin at the point y in the direction $x-y$. If $K(y)$ is a cone with its vertex at y, then the set $K=K(y)$ is a cone. One can easily show that a cone K is convex if and only if, for any two points x and y in it, it includes also their sum $x+y$.

Examples of convex cones are linear subsets of X and also cones consisting of a single ray $\{\alpha u\}_{\alpha \geqslant 0}$, where $u \in X$. The set

$$\{\alpha u\}_{\alpha \geqslant 0} \cup \{\beta v\}_{\beta \geqslant 0}$$

where u and v are not null elements of X and $u \neq \pm v$, is a nonconvex cone. In the space $C(E)$, the set of functions x that are nonnegative at every point $t \in E$ is a convex corporeal cone. In the space $L^p(E)$ where $1 \leqslant p \leqslant \infty$, a convex cone defines functions that are nonnegative almost everywhere. If $p < \infty$, this cone is not corporeal.

Suppose that $\Omega \subset X$. We shall call the set

$$\{y \in X \mid y = \alpha x,\ \alpha > 0,\ x \in \Omega\}$$

the ***conical envelope*** of Ω and shall denote it by $C(\Omega)$. We can show that $C(\Omega)$ is the smallest cone containing Ω; in other words, if a cone K contains Ω, it also contains $C(\Omega)$. In what follows, an important role will be played by the concept of a dual cone. Let K denote a cone contained in X. Let us define

$$K^* = \{f \in X^* \mid f(x) \geqslant 0 \quad \text{for every} \quad x \in K\}.$$

One can easily show that K^* is a convex closed cone in X^*. This cone is called the ***dual*** to K (or the ***polar cone*** to K).

Let us give some examples. If K is a linear set, then K^* consists of all functionals that vanish on the set K. Thus, in this case, K^* is a subspace of X. The cone K^* dual to the cone K of all ***nonnegative*** functions in the space $C(E)$ consists of all nonnegative measures $\mu \in V(E)$. Finally, the cone dual to the cone of all non-negative functions in $L^p(E)$, where $1 \leqslant p < \infty$, consists of all non-negative functions belonging to $L^q(E)$, where $1/p + 1/q = 1$.

We shall refer to a set N as a ***planar set*** if, for any two of its points, it contains the straight line passing through those points (in other words, for $x, y \in N$ and any real α, we have $\alpha x + (1-\alpha) y \in N$. Obviously, a planar set containing 0 is linear. The ***planar hull*** of a set $\Omega \subset X$ is defined as the intersection of all planar sets containing Ω. The ***linear hull*** of the set Ω is the intersection of all linear sets containing Ω. We shall denote the linear hull of Ω by $L(\Omega)$. One can easily show that

$$L(\Omega) = \left\{ y \in X \mid y = \sum_{k=1}^{n} \alpha_k x_k;\ n = 1, 2, \ldots;\ x_1, \ldots, x_n \in \Omega \right\}.$$

The ***hyperplane*** H of the space X is defined as the maximal planar set that does not coincide with X. In other words, the hyperplane H is a planar set and if H' is a planar set containing H, then either $H' = X$ or $H' = H$. It can be shown that, if $f \in X^*$ and c is a real number, then the set

$$f^{-1}(c) = \{ y \in X \mid f(y) = c \}$$

is a closed hyperplane. Conversely, for any closed hyperplane H, there exists a functional $f \in X^*$ and a real number c such that $H = f^{-1}(c)$. Suppose that $f \in X^*$ and $H = f^{-1}(0)$ and let y denote an element of X such that $f(y) = 1$: Then, as one can easily see, every element x of X can be represented in one and only one way in the form $x = f(x)y + h$, where $h \in H$. We note also that, if H is a closed hyperplane including zero and if f and g are two functionals belonging to X^* such that $H = f^{-1}(0)$ and $H = g^{-1}(0)$, then f is proportional to g [that is, there exists a nonzero number λ such that $f(x) \equiv \lambda g(x)$].

Let H denote a closed hyperplane in X and let c denote a real number such that $H = f^{-1}(c)$. Then, the sets

$$R' = f^{-1}((c, \infty)) = \{ y \in X \mid f(y) > c \}$$

and

$$R'' = f^{-1}((-\infty, c)) = \{ y \in X \mid f(y) < c \}$$

are called the ***open half-spaces generated by the hyperplane*** H. The sets $Q' = f^{-1}([c, \infty))$ and $Q'' = f^{-1}((-\infty, c])$ are called the ***closed half-spaces generated*** by H. We note that both closed and open half-spaces are convex cones K with vertex at an arbitrary point in H.

3°. We shall say that a sequence $\{x_n\}$ of elements of a normed space X ***converges weakly*** to an element $x \in X$ and we shall write this

$$x_n \xrightarrow{w} x$$

if $f(x_n) \to f(x)$ for every $f \in X^*$. We mention that the relation $x_n \xrightarrow{w} x$ implies that

$$\|x\| \leqslant \varliminf_{n \to \infty} \|x_n\|.$$

A subset Ω of X is said to be ***weakly sequentially closed*** (weakly closed with respect to sequences) if the relation $x_n \xrightarrow{w} x$, where each $x_n \in \Omega$ implies that $x \in \Omega$. We mention that a closed set is not necessarily weakly sequentially closed. However, in reflexive spaces,[1] in particular in $L^p(E)$, where $1 < p < \infty$, a convex closed set is weakly sequentially closed.

A subset Ω of X is said to be ***compact*** if every sequence of elements x_n of Ω contains a subsequence $\{x_{n_{ij}}\}$ that converges (with respect to the norm) to an element of Ω. Let X denote a Banach space. A subset Ω of X is said to be ***weakly compact*** if every sequence of elements x_n of Ω contains a sequence that converges weakly to an element of Ω. We note that weakly compact and ***a fortiori*** compact sets are bounded. If every closed bounded subset of a space X is compact, then X is a finite-dimensional space. In reflexive spaces, every bounded weakly sequentially closed set is weakly compact.

Weak convergence can also be defined for linear functionals. A sequence $\{f_n\}$ of linear functionals defined on a normed space X is said to ***converge weakly*** to a functional $f \in X^*$ and we denote this by writing

$$f_n \xrightarrow{w} f$$

[1] A space X is said to be *reflexive* if $X^{**} = X$.

if $f_n(x) \to f(x)$ for every $x \in X$. One can show that the relation $f_n \xrightarrow{w} f$ implies

$$\|f\| \leqslant \varliminf_{n\to\infty} \|f_n\|.$$

A subset U of X^* is said to be ***weakly sequentially closed*** if the relation $f_n \xrightarrow{w} f$, where each $f_n \in U$, implies $f \in \Omega$. A subset U of X^* is said to be ***weakly compact*** if an arbitrary sequence $\{f_n\}$ of members of U contains a subsequence $\{f_{n_i}\}$ that converges weakly to a member of U. It is known (see, for example, Theorem 2 (I.VIII) in [32]) that, if X is a separable space,[1] then every bounded weakly sequentially closed subset of X^* is weakly compact.

4°. Let X and Y denote normed spaces. Let Ω denote a subset of X and let A denote an operator mapping Ω into Y. In the sequel, unless the contrary is stated, we shall assume that the domain of definition of the operator A (the set Ω) coincides with the entire space X.

An operator A is said to be ***continuous at a point*** x is the relation $x_n \to x$ implies $Ax_n \to Ax$ and it is said to be ***continuous on a set*** Ω if it is continuous at every point of Ω. In what follows, when we speak of a continuous operator, we shall usually mean an operator that is continuous at every point of its domain of definition. An operator A is said to be ***bounded*** if it maps every bounded set into a bounded set.

We point out that continuity and boundedness of A are not in general equivalent. However, if A is an additive operator (that is, $A(x+y) = Ax + Ay$), then continuity of A is equivalent to its boundedness. An additive continuous operator is said to be ***linear***. We note that linearity of an operator implies its homogeneity. We shall usually use the notation $[X \to Y]$ to indicate the normed space of all linear operators that map X into Y with norm defined by

$$\|A\| = \sup_{\|x\| \leqslant 1} \|Ax\|.$$

[1] A space is said to be *separable* if it contains an everywhere-dense countable subset. The spaces $C(E)$ and $L^p(E)$, where $(1 \leqslant p < \infty)$, are separable.

An operator that maps a subset Ω of X into the set of real numbers is called a ***functional***. A functional f is said to be ***weakly continuous*** at a point $x \in X$ if for any sequence of members of X that converges weakly to x, the sequence $\{f(x_n)\}$ converges to $f(x)$. A functional f is said to be ***lower-semicontinuous*** (resp. ***weakly lower-semicontinuous***) at a point x if the relation $x_n \to x$ (resp. the relation $x_n \xrightarrow{w} x$) implies

$$f(x) \leqslant \varliminf_{n \to \infty} f(x_n).$$

Finally, a functional f is said to be ***upper-semicontinuous*** (resp. ***weakly upper-semicontinuous***) at x if

$$f(x) \geqslant \varlimsup_{n \to \infty} f(x_n)$$

for an arbitrary sequence $\{x_n\}$ that converges weakly to x. We note that continuity (resp. weak continuity) of a functional at a point implies its semicontinuity (resp. weak semicontinuity), both upper and lower, at that point. Conversely, if a functional is upper- and lower-semicontinuous (resp. weakly upper- and lower-semicontinuous), it is continuous (resp. weakly continuous). If a functional f is lower-semicontinuous (resp. weakly lower-semicontinuous), the function $-f$ is upper-semicontinuous (resp. weakly upper-semicontinuous) and vice versa. Therefore, it will be sufficient for us to study the properties of lower-semicontinuous and weakly lower-semicontinuous functionals. We shall say that a functional is weakly continuous or semicontinuous or weakly semicontinuous (either upper- or lower-) on a set if it possesses the property in question at every point of that set.

If a functional f satisfies the first of the inequalities

$$\begin{gathered} f\left(\frac{1}{2}x + \frac{1}{2}y\right) \leqslant \max\{f(x), f(y)\} \\ f\left(\frac{1}{2}x + \frac{1}{2}y\right) \geqslant \min\{f(x), f(y)\} \end{gathered} \tag{1.3}$$

it is said to be ***quasiconvex***; if it satisfies the second of these inequalities, it is said to be ***quasiconcave***. In the case of strict inequality in (1.3), the functional f is said to be ***strictly quasiconvex*** or ***strictly quasi-concave***.

A functional f is said to be convex if the first of the inequalities

$$f\left(\frac{1}{2}x+\frac{1}{2}y\right)\leqslant\frac{1}{2}(f(x)+f(y))$$
$$f\left(\frac{1}{2}x+\frac{1}{2}y\right)\geqslant\frac{1}{2}(f(x)+f(y)) \tag{1.4}$$

is satisfied; it is said to be ***concave*** if the second inequality is satisfied. Again, if strict inequality holds, the functional f is said to be ***strictly convex*** or ***strictly concave***. We note that quasiconvexity (resp. convexity) of f implies quasiconcavity (resp. concavity) of the functional $-f$ and vice versa. Therefore, it will be sufficient for us to study quasiconvex and convex functionals. It is easy to show that quasiconvexity of a functional f is equivalent to convexity of the set

$$\{y\in\Omega\,|\,f(y)\leqslant f(x)\}$$

for arbitrary $x\in X$. We note also that (strict) convexity of a functional implies its (strict) quasiconvexity, although the converse of this is not true.

In what follows, we shall often consider only continuous quasiconvex (and hence convex) functionals–often without explicitly stating this. We mention that continuity of a quasiconvex functional implies that the first of the inequalities

$$f(\alpha x+(1-\alpha)y)\leqslant\max\{f(x),\ f(y)\}$$
$$(f(\alpha x+(1-\alpha)y)\leqslant\alpha f(x)+(1-\alpha)f(y)) \tag{1.5}$$

holds and that continuity of a convex functional implies that the second of these inequalities holds. Here, $\alpha\in(0,\ 1)$. These inequalities will be strict in the case of continuity of a strictly quasiconvex or a strictly convex functional [7]. The following theorem will be important [44]:

THEOREM 1.1. *A quasiconvex functional defined on a Banach space X is weakly lower-semicontinuous.*

Let us look at an example. Let E denote a measurable subset of n-dimensional space. Consider a function of two variables $g\,(x\,(t)$

defined for $-\infty < x < \infty$ and $t \in E$ that satisfies the following three conditions:

(a) $g(x, t)$ is continuous with respect to x for almost all $t \in E$,
(b) $g(x, t)$ is measurable with respect to t for all $x \in (-\infty, \infty)$,
(c) there exist numbers p_1 and p_2 in $(1 \leqslant p_i < \infty,\ i=1, 2)$ such that

$$|g(x, t)| \leqslant \alpha(t) + b|x|^{p_1/p_2},$$

where b is a positive constant and $\alpha(t) \in L^{p_2}(E)$. Under these assumptions the operator A defined by

$$Ax(t) = g(x(t), t)$$

maps $L^{p_1}(E)$ into $L^{p_2}(E)$ and it is continuous (see [36, Chapter 1, Sec. 2]. The operator A is sometimes called "Nemytskiy's operator."

Let h denote a linear functional defined on $L^{p_2}(E)$; that is, suppose that

$$h(y) = \int_E y(t)h(t)\,dt, \quad h(t) \in L^{q_2}(E), \quad 1/p_2 + 1/q_2 = 1.$$

Consider the functional f defined on $L^{p_1}(E)$ by

$$f(x) = \int_E g(x(t), t)\,h(t)\,dt.$$

Obviously, f is a continuous functional. Let us suppose now that the function g is convex with respect to x for almost all t, that is, that

$$g\left(\frac{x_1+x_2}{2}, t\right) \leqslant \frac{1}{2}[g(x_1, t) + g(x_2, t)]$$

for almost all t and that h is almost everywhere nonnegative. Then, as one can easily see, f is a convex functional. In this case, it follows from Theorem 1.1 that the functional f is weakly lower-semicontinuous.

2. Sublinear functionals.

1°. A functional p defined on a linear set X is said to be *semi-additive* (or *subadditive*) if

$$p(x+y) \leqslant p(x)+p(y)$$

for all x and y in X.

A functional p is said to be ***positively homogeneous*** if, for every $\lambda > 0$,

$$p(\lambda x) = \lambda p(x).$$

Let us point out a few properties enjoyed by positively homogeneous semiadditive functionals:

$$\text{a)} \quad p(0) = 0. \tag{2.1}$$

This follows from the fact that

$$p(0) = p(2 \cdot 0) = 2p(0).$$

$$\text{b)} \quad -p(-x) \leqslant p(x). \tag{2.2}$$

This is true because $0 = p(0) = p(x-x) \leqslant p(x) + p(-x)$, which is equivalent to (2.2). One consequence of (2.2) is the fact that a functional $p \neq 0$ must assume positive values at certain points of the set X. (if $p(-x) < 0$, then $p(x) \geqslant -p(-x) > 0$.) On the other hand, there exist positively homogeneous semiadditive functionals that never assume negative values (for example, the functional $p(x) = \|x\|$ constituting the norm in a normed space).

In what follows, we shall assume that X is a normed space and we shall be interested only in continuous positively homogeneous semiadditive functionals. We shall say that such functionals are ***sublinear functionals***.[1]

One can easily show that a positively homogeneous semiadditive functional p that is continuous at 0 is continuous at an arbitrary point $x \in X$ and hence that it is sublinear. To see this, suppose that $x_n \to x$. Then,

$$p(x_n) = p((x_n - x) + x) \leqslant p(x_n - x) + p(x).$$

Since $x_n - x \to 0$, the continuity of p at 0 implies that $p(x_n - x) \to 0$.

[1]In [25], such functionals are called *linearly convex*.

By using this fact and taking the limit in the preceding inequality, we have

$$\overline{\lim}\, p(x_n) \leqslant p(x).$$

Similarly, from the inequality

$$p(x) = p((x - x_n) + x_n) \leqslant p(x - x_n) + p(x_n)$$

we obtain the result that

$$p(x) \leqslant \underline{\lim}\, p(x_n).$$

Thus,

$$p(x) \leqslant \underline{\lim}\, p(x_n) \leqslant \overline{\lim}\, p(x_n) \leqslant p(x),$$

from which it follows that p is continuous at the point x.

A positively homogeneous semiadditive functional p is said to be *bounded* if there exists a constant c such that $|p(x)| \leqslant c\|x\|$ for all $x \in X$.

THEOREM 2.1. *For a positively homogeneous semiadditive functional to be sublinear, it is necessary and sufficient that it be bounded.*

The proof of this theorem coincides with the proof of the analogous theorem for linear operators (see [32, Theorem I.III]). If a functional is sublinear, then

$$\|p\| = \sup_{\|x\| \leqslant 1} |p(x)| < \infty.$$

We shall call the quantity $\|p\|$ the *norm* of the functional p. One can easily show that, for arbitrary $x \in X$,

$$|p(x)| \leqslant \|p\| \cdot \|x\|. \tag{2.3}$$

Here, if c is a constant such that $|p(x)| \leqslant c\|x\|$, then $\|p\| \leqslant c$.

Let us prove the validity of the inequality (important in what follows)

$$|p(x_1)-p(x_2)| \leqslant \|p\| \cdot \|x_1 - x_2\| \tag{2.4}$$

for any x_1 and x_2 in X. Since p is sublinear, it follows that

$$p(x_1)=p(x_1-x_2+x_2) \leqslant p(x_1-x_2)+p(x_2),$$
$$p(x_2)=p(x_2-x_1+x_1) \leqslant p(x_2-x_1)+p(x_1).$$

By transforming these inequalities, we obtain

$$p(x_1)-p(x_2) \leqslant p(x_1-x_2) \leqslant |p(x_1-x_2)| \leqslant \|p\| \cdot \|x_1-x_2\|,$$
$$p(x_2)-p(x_1) \leqslant p(x_2-x_1) \leqslant |p(x_2-x_1)| \leqslant \|p\| \cdot \|x_2-x_1\|,$$

from which (2.4) follows.

The set P of all functionals that are sublinear in a space X possesses the following properties:

1) Suppose that $q(u_1, u_2, \ldots, u_n)$ is a nondecreasing continuous positively homogeneous subadditive function of n variables (in other words, a nondecreasing sublinear functional in the space R^n). Then, the functional p defined by

$$p(x)=q(p_1(x), p_2(x), \ldots, p_n(x)),$$

where $p_i \in P$ for $(i=1,2,\ldots, n)$, is also an element of P.

To see this, let x and y denote members of X and let λ denote a nonnegative real number. Then,

$$p(x+y)=q(p_1(x+y), p_2(x+y), \ldots, p_n(x+y)) \leqslant$$
$$\leqslant q(p_1(x)+p_1(y), p_2(x)+p_2(y), \ldots, p_n(x)+p_n(y)) \leqslant$$
$$\leqslant q(p_1(x), p_2(x), \ldots, p_n(x))+q(p_1(y), p_2(y), \ldots, p_n(y))=$$
$$=p(x)+p(y),$$
$$p(\lambda x)=q(\lambda p_1(x), \lambda p_2(x), \ldots, \lambda p_n(x))=$$
$$=\lambda q(p_1(x), \ldots, p_n(x))=\lambda p(x).$$

Thus, p is positively homogeneous and semiadditive. The continuity of p is obvious. In particular, for $q(u_1, u_2)=\lambda u_1+\mu u_2$, where λ and μ are nonnegative numbers, we obtain the result that $\lambda p_1+\mu p_2$ belong to P whenever p_1 and p_2 do. Here, as one can easily see,

$$\|\lambda p_1+\mu p_2\| \leqslant \lambda \|p_1\|+\mu \|p_2\|.$$

2) Let A denote a set and for every α in A let p_α denote a member of P. Suppose that

$$\sup_{\alpha\in A}\|p_\alpha\|<\infty.$$

Then, as one can easily see, the functional p defined by

$$p(x)=\sup_{\alpha\in A}p_\alpha(x)$$

is sublinear.

We mention the validity of the following assertion: Let A denote a linear operator defined from X into Y and let p denote a sublinear functional in the space Y. Then the functional p_A defined in the space X by $p_A(x)=p(Ax)$ is sublinear.

Let us look at some examples of sublinear functionals.

1. Obviously, the functional $p(x)=\|x\|$ is sublinear.

2. If $f\in X^*$, then f is sublinear. The functional $p(x)=|f(x)|$ is also sublinear.

3. Suppose $X=C(E)$, where E is a closed bounded subset of a finite-dimensional space. Let E_0 denote a closed subset of E. One can easily show that the functionals

$$p_1(x)=\max_{t\in E_0}x(t)\ p_2(x)=\max_{t\in E_0}|x(t)|$$

are sublinear.

4. Suppose that $X=L^\infty(E)$, where E is a measurable subset of a finite-dimensional space and let E_0 denote a subset of E. Then the functionals

$$p_1(x)=\operatorname{vrai}\max_{t\in E_0}x(t),\quad p_2=\operatorname{vrai}\max_{t\in E_0}|x(t)|$$

are sublinear

5. In the space $L^1(E)$ [where E is as in the preceding example],

$$p(x)=\int_{E_0}(|x(t)|+a(t)x(t))\,dt,$$

the functional where $E_0\subset E$ and $a\in L^\infty(E_0)$, is a sublinear functional.

We shall call a semiadditive continuous functional a *seminorm* if, for every real number λ,

$$p(\lambda x)=|\lambda|p(x).$$

Obviously, a seminorm is a sublinear functional. Examples of seminorms are the norm, the functional p in example 2), the functionals p_2 in examples 3) and 4), and the functional p in example 5) when $a = 0$. It follows from the inequality

$$p(x) = \frac{1}{2}(p(x) + p(-x)) \geqslant \frac{1}{2} p(x + (-x)) = p(0) = 0$$

that a seminorm assumes only nonnegative values. We mention that, in contrast with a norm, the equation $p(x) = 0$ (where p is a seminorm) does not imply that $x = 0$. A seminorm does, however, satisfy all the other axioms of a norm.

2°. We shall call a linear functional f a *supporting* functional for a sublinear functional p if, for all $x \in X$,

$$f(x) \leqslant p(x).$$

We denote by $\mathfrak{u}_p$ the set of all supporting functionals for p. First of all, let us show that, for any $p \in P$, the set $\mathfrak{u}_p \neq \varnothing$. To do this, we need to use one of the fundamental theorems of functional analysis, namely, the Hahn-Banach theorem (see [32, Theorem 7 (2.IV)]).

THE HAHN-BANACH THEOREM. *Let X denote a linear set. Let p denote a positively homogeneous semiadditive functional defined on X and let f_0 denote an additive homogeneous function defined on a linear set $X_0 \subset X$. Suppose that $f_0(x) \leqslant p(x)$ everywhere on X_0. Then, there exists an additive homogeneous functional f defined on all X that coincides with f_0 on X_0 and satisfies X the condition $f(x) \leqslant p(x)$.*

THEOREM 2.2. *If $p \in P$, then, for every $x_0 \in X$, there exists an $f \in X^*$ such that $f(x_0) = p(x_0)$ and $f(x) \leqslant p(x)$ for $x \in X$.*

Proof: Let us take $X_0 = \{\lambda x_0 \mid \lambda \in (-\infty, \infty)\}$ and let us define on X_0 a linear functional f_0 by

$$f_0(\lambda x_0) = \lambda p(x_0).$$

Let us show that, for all $x \in X_0$,

$$f_0(x) \leqslant p(x).$$

If $\lambda \geqslant 0$, we have $f_0(\lambda x_0) = \lambda f_0(x_0) = \lambda p(x_0) = p(\lambda x_0)$. On the other hand, if $\lambda < 0$, then, by virtue of (2.2), we have

$$p(\lambda x_0) = |\lambda| p(-x_0) \geqslant -|\lambda| p(x_0) = \lambda f_0(x_0) = f_0(\lambda x_0).$$

On the basis of the Hahn-Banach theorem, there exists an additive homogeneous functional f that coincides with f_0 on X_0 [that is, such that $f(x_0) = f_0(x_0) = p(x_0)$] and satisfies on X the inequality

$$f(x) \leqslant p(x). \tag{2.5}$$

Let us show that $f(x) \geqslant -p(-x)$. It follows from (2.5) that $f(-x) \leqslant p(-x)$, so that $f(x) = -f(-x) \geqslant -p(-x)$. Thus [cf. (2.2)],

$$-p(-x) \leqslant f(x) \leqslant p(x). \tag{2.6}$$

From this inequality it follows that the functional f is continuous at 0, that is, that $f \in X^*$. This completes the proof of the theorem.

For $x \in X$, let us define

$$\boldsymbol{u}_p^x = \{f \in \boldsymbol{u}_p \,|\, f(x) = p(x)\}.$$

It follows from Theorem 2.2 that the set $\boldsymbol{u}_p^x \neq \varnothing$ for every $x \in X$, so that *a fortiori* the set $\boldsymbol{u}_p$ is nonempty.

One can easily show that $\boldsymbol{u}_p$ is a convex bounded weakly sequentially closed set in the space X^*. It follows from Theorem 2.2 that

$$p(x) = \max_{f \in \boldsymbol{u}_p} f(x). \tag{2.7}$$

In what follows, we shall assume that X is separable.[1] Then (see Sec. 1, 3°), $\boldsymbol{u}_p$ is a weakly compact set. Let us denote by M the set of all convex weakly compact subsets of X^*. To every $\boldsymbol{u} \in M$, we can assign a sublinear functional $p_{\boldsymbol{u}}$:

[1]We made the assumption of separability only to make the proofs easier. One can show that all the theorems given in the present section remain valid in the case of an arbitrary locally convex space [55].

$$p_{\boldsymbol{u}}(x) = \max_{f \in \boldsymbol{u}} f(x). \tag{2.8}$$

Formula (2.8) defines an operator $\varphi : \varphi(\boldsymbol{u}) = p_{\boldsymbol{u}}$ that maps the set M into the set P of all sublinear functionals defined on X. We have the following theorem:

THEOREM 2.3. *The operator φ defines a one-to-one correspondence between M and P; in other words, for every $p \in P$ there exists a unique $\boldsymbol{u} \in M$ such that $p = \varphi(\boldsymbol{u})$.*

Proof:[1] It follows immediately from (2.7) that, for any $p \in P$, we have $p = \varphi(\boldsymbol{u}_p)$. Let us show now that if $\boldsymbol{u} \neq \boldsymbol{v}$, then $\varphi(\boldsymbol{u}) \neq \varphi(\boldsymbol{v})$. Without loss of generality, we may assume that there exists an $f' \in X^*$ such that $f' \in \boldsymbol{u}$ but $f' \notin \boldsymbol{v}$. Since $\boldsymbol{v}$ is closed in the topology $\sigma(X^*, X)$ defined by the duality between X and X^* and since (in accordance with a familiar theorem) a closed convex set in a locally convex space can be separated from any point not belonging to it, there exists a point $x \in X$ such that

$$f'(x) > \max_{f \in \boldsymbol{v}} f(x).$$

But then *a fortiori*

$$\varphi(\boldsymbol{u})(x) = \max_{f \in \boldsymbol{u}} f(x) \geqslant f'(x) > \max_{f \in \boldsymbol{v}} f(x) = \varphi(\boldsymbol{v})(x).$$

from which it follows that $\varphi(\boldsymbol{u}) \neq \varphi(\boldsymbol{v})$. This completes the proof of the theorem.

It follows from Theorem 2.3 that the operator φ has an inverse operator φ^{-1}. Therefore,

$$\varphi^{-1}(p) = \boldsymbol{u}_p. \tag{2.9}$$

In the set M, let us define operations of addition and multiplication by a nonnegative number as follows:

[1]The reader who is unfamiliar with the theory of locally convex spaces can omit this proof without any loss in understanding the sequel.

$$\boldsymbol{u}+\boldsymbol{v}=\{f\in X^* \mid f=g+h;\ g\in\boldsymbol{u},\ h\in\boldsymbol{v}\},$$
$$\lambda\boldsymbol{u}=\{f\in X^* \mid f=\lambda g;\ g\in\boldsymbol{u}\}.$$

where $\boldsymbol{u}$ and $\boldsymbol{v}$ are members of M and $\lambda\geqslant 0$. By using the weak compactness of $\boldsymbol{u}$ and $\boldsymbol{v}$, one can easily show that $\boldsymbol{u}+\boldsymbol{v}$ and $\lambda\boldsymbol{u}$ belong to M. These algebraic operations obviously possess the usual properties:

$$\boldsymbol{u}+\boldsymbol{v}=\boldsymbol{v}+\boldsymbol{u};\quad (\boldsymbol{u}+\boldsymbol{v})+\boldsymbol{w}=\boldsymbol{u}+(\boldsymbol{v}+\boldsymbol{w});\quad 0\cdot\boldsymbol{u}=\mathbf{0}$$

$$=\lambda\boldsymbol{u}+\lambda\boldsymbol{v};\quad (\lambda+\mu)\,\boldsymbol{u}=\lambda\boldsymbol{u}+\mu\boldsymbol{u};\quad \lambda(\mu\boldsymbol{u})=(\lambda\mu)\,\boldsymbol{u};\quad 1\cdot\boldsymbol{u}=\boldsymbol{u}.$$

(Here, $\mathbf{0}$ is the set whose only member is the zero element of X^*.) We shall assume that M is partially ordered by inclusion.

THEOREM 2.4. *The operator* φ *possesses the following properties: for* $\boldsymbol{u}$, $\boldsymbol{v}\in M$,

1) $\varphi(\boldsymbol{u}+\boldsymbol{v})=\varphi(\boldsymbol{u})+\varphi(\boldsymbol{v})$,

2) $\varphi(\lambda\boldsymbol{u})=\lambda\varphi(\boldsymbol{u})$ $(\lambda\geqslant 0)$,

3) *for arbitrary* $x\in X$, *the inequality* $\varphi(\boldsymbol{u})(x)\geqslant\varphi(\boldsymbol{v})(x)$ *holds if and only if* $\boldsymbol{u}\supset\boldsymbol{v}$.

Proof: 1) Let x denote a member of X and let f' and g' denote functionals in X^* such that

$$\varphi(\boldsymbol{u})(x)=\max_{f\in\boldsymbol{u}} f(x)=f'(x),$$
$$\varphi(\boldsymbol{v})(x)=\max_{g\in\boldsymbol{v}} g(x)=g'(x).$$

We have

$$\varphi(\boldsymbol{u}+\boldsymbol{v})(x)=\max_{h\in\boldsymbol{u}+\boldsymbol{v}} h(x)=\max_{f\in\boldsymbol{u},\,g\in\boldsymbol{v}}(f(x)+g(x))\geqslant f'(x)+g'(x)=$$
$$=\varphi(\boldsymbol{u})(x)+\varphi(\boldsymbol{v})(x).$$

On the other hand,

$$\varphi(\boldsymbol{u}+\boldsymbol{v})(x)=\max_{f\in\boldsymbol{u},\,g\in\boldsymbol{v}}(f(x)+g(x))\leqslant\max_{f\in\boldsymbol{u}} f(x)+\max_{g\in\boldsymbol{v}} g(x)=$$
$$=\varphi(\boldsymbol{u})(x)+\varphi(\boldsymbol{v})(x).$$

Thus,

$$\varphi(\boldsymbol{u}+\boldsymbol{v})(x)=\varphi(\boldsymbol{u})(x)+\varphi(\boldsymbol{v})(x).$$

2) For any x in X and any $\lambda \geqslant 0$,

$$\varphi(\lambda \boldsymbol{u})(x)=\max_{f\in\lambda \boldsymbol{u}} f(x)=\max_{g\in \boldsymbol{u}} \lambda g(x)=\lambda \max_{g\in \boldsymbol{u}} g(x)=\lambda\varphi(\boldsymbol{u})(x).$$

3) If $\boldsymbol{u}\supset\boldsymbol{v}$, then

$$\max_{f\in \boldsymbol{u}} f(x)\geqslant \max_{f\in \boldsymbol{v}} f(x)$$

for every $x\in X$. Conversely, if $\varphi(\boldsymbol{u})(x)\geqslant\varphi(\boldsymbol{v})(x)$ for every x in X, then every supporting functional for $\varphi(\boldsymbol{v})$ is also a supporting functional for $\varphi(\boldsymbol{u})$. Since the set of all supporting functionals for $\varphi(\boldsymbol{u})$ coincides with $\boldsymbol{u}$ and the set of all supporting functionals $\varphi(\boldsymbol{v})$ coincides with $\boldsymbol{v}$, we have $\boldsymbol{u}\supset\boldsymbol{v}$. This completes the proof of the theorem.

THEOREM 2.5.[1] *Let p_1 and p_2 denote members of P and let λ_1 and λ_2 denote nonnegative numbers. Define $p=\lambda_1 p_1+\lambda_2 p_2$. Then, $\boldsymbol{u}_p=\lambda_1\boldsymbol{u}_{p_1}+\lambda_2\boldsymbol{u}_{p_2}$.*

Proof: Since $p_1=\varphi(\boldsymbol{u}_{p_1})$, $p_2=\varphi(\boldsymbol{u}_{p_2})$, it follows on the basis of Theorem 2.4 that

$$\lambda_1 p_1+\lambda_2 p_2=\varphi(\lambda_1\boldsymbol{u}_{p_1}+\lambda_2\boldsymbol{u}_{p_2}).$$

Therefore,

$$\boldsymbol{u}_p=\varphi^{-1}(p)=\varphi^{-1}(\lambda_1 p_1+\lambda_2 p_2)=\varphi^{-1}(\varphi(\lambda_1\boldsymbol{u}_{p_1}+\lambda_2\boldsymbol{u}_{p_2}))=\lambda_1\boldsymbol{u}_{p_1}+\lambda_2\boldsymbol{u}_{p_2}.$$

This completes the proof of the theorem.

Let L denote a bounded subset of X^*. The smallest convex weakly compact set containing L, that is, the intersection of all convex weakly compact sets containing L will be denoted by $\overline{\text{co}}\,(L)$.

[1]This theorem was given without proof in [25].

THEOREM 2.6. *Let A denote a set and, for every* $\alpha \in A$, *let* p_α *denote an element of P. Suppose that*

$$\sup_{\alpha \in A} \|p_\alpha\| < \infty$$

Denote by p the functional defined by

$$p(x) = \sup_{\alpha \in A} p_\alpha(x).$$

Then,

$$\mathfrak{u}_p = \overline{\text{co}}\left(\bigcup_{\alpha \in A} \mathfrak{u}_{p_\alpha}\right).$$

Proof: Since $p(x) \geqslant p_\alpha(x)$ for every α in A and every x in X, it follows on the basis of Theorem 2.4 that $\mathfrak{u}_p \supset \mathfrak{u}_{p_\alpha}$ for every α in A. But then

$$\mathfrak{u}_p \supset \bigcup_{\alpha \in A} \mathfrak{u}_{p_\alpha}$$

and, since $\mathfrak{u}_p$ is weakly compact and convex,

$$\mathfrak{u}_p \supset \overline{\text{co}}\left(\bigcup_{\alpha \in A} \mathfrak{u}_{p_\alpha}\right).$$

The set

$$\mathfrak{v} = \overline{\text{co}}\left(\bigcup_{\alpha \in A} \mathfrak{u}_{p_\alpha}\right)$$

is a member of M. Let us suppose that $\mathfrak{u}_p \neq \mathfrak{v}$. Then,

$$p = \varphi(\mathfrak{u}_p) \neq \varphi(\mathfrak{v}) = p_{\mathfrak{v}}$$

and, since $\mathfrak{u}_p \supset \mathfrak{v}$, there exists an x_0 in X such that $p(x) > p_v(x_0)$. On the other hand, $p_v(x_0) \geqslant p_\alpha(x_0)$ for all a (since $\mathfrak{v} \supset \mathfrak{u}_{p_\alpha}$) and therefore we also have

$$p_v(x_0) \geqslant \sup_\alpha p_\alpha(x_0) = p(x_0).$$

This contradiction proves the theorem.

The relationship between the sets P and M is treated in greater detail in [55].

3°. Let us see how the set $\mathfrak{u}_p$ is defined for certain classes of sublinear functionals.

1) Let f denote a sublinear functional belonging to X^*. Here, as one can easily see, $\mathfrak{u}_f = \{f\}$.

2) Suppose that $p(x) = \|x\|$. In this case,

$$\mathfrak{u}_p = \{f \in X^* \mid \|f\| \leqslant 1\}. \tag{2.10}$$

To see this, note that, if $f \in \mathfrak{u}_p$, then, for arbitrary x,

$$f(x) \leqslant \|x\|, \quad -f(x) = f(-x) \leqslant \|-x\| = \|x\|.$$

Therefore, $|f(x)| \leqslant \|x\|$ and

$$\|f\| = \sup_{\|x\| \leqslant 1} |f(x)| \leqslant 1.$$

Conversely, if $\|f\| \leqslant 1$, then

$$f(x) \leqslant |f(x)| \leqslant \|f\| \cdot \|x\| \leqslant \|x\|.$$

3) Let p denote a seminorm. Reasoning as in the preceding example, we can easily show that

$$\mathfrak{u}_p = \{f \in X^* \mid \|f\|_p \leqslant 1\}, \tag{2.11}$$

where $\|f\|_p = \sup_{p(x) < 1} |f(x)|$.

Suppose that $X_0 = \{x \in X \mid p(x) = 0\}$. It follows from the properties of a seminorm that X_0 is a closed linear set, that is, a subspace of X. It follows from the definition of $\|f\|_p$ that the functional $f \in \mathfrak{u}_p$ has the property that $f(X_0) = 0$.

4°. Let us look at some examples.

1) Suppose that $X = C(E)$, where E is a closed bounded subset of a finite-dimensional space, and

$$p(x) = \max_{t \in E} |x(t)|.$$

Since $p(x)=\|x\|$ and since the dual to $C(E)$ can be identified with the space $V(E)$ of completely additive measures defined on E, we can, on the basis of (2.10) assert that

$$u_p=\{\mu\in V(E)\,|\,\|\mu\|=\operatorname{var}\mu(E)\leqslant 1\}.$$

2) As before, suppose that $X=C(E)$ Let E_0 denote a closed subset of E and take

$$p(x)=\max_{t\in E_0}|x(t)|.$$

Obviously, p is a seminorm in $C(E)$. One can easily show that if the functional f is generated by the measure μ, then the inequality

$$\|f\|_p=\sup_{p(x)\leqslant 1}|f(x)|\leqslant 1$$

holds if and only if

$$\operatorname{var}\mu(E_0)\leqslant 1,\ \ \operatorname{var}\mu(E\setminus E_0)=0.$$

By using (2.11), we obtain

$$u_p=\{\mu\in V(E)\,|\operatorname{var}\mu(E_0)\leqslant 1,\ \ \operatorname{var}\mu(E\setminus E_0)=0\}.$$

3) Again, let $X=C(E)$, let E_0 denote a closed subset of E, and take

$$p(x)=\max_{t\in E_0}x(t).$$

Let t denote a member of E_0. Consider the linear functional generated by the measure μ_t: for $e\subset E$,

$$\mu_t(e)=\begin{cases}1, & t\in e,\\ 0, & t\notin e.\end{cases}$$

We shall also denote this functional by μ_t. The measure μ_t has the following properties:

1) $\mu_t(e)\geqslant 0$ for every $e\subset E$ $(\mu_t\geqslant 0)$;
2) $\mu_t(E_0)=\operatorname{var}\mu_t(E_0)=1$;

3) $\mu_t(E \setminus E_0) = 0$;

4) $\int_E x d\mu_t = \mu_t(x) = x(t)$ for every $x \in C(E)$.

It follows from the last property that

$$p(x) = \max_{t \in E_0} x(t) = \max_{t \in E_0} \mu_t(x).$$

Since μ_t is a linear functional, we have $u_{\mu_t} = \{\mu_t\}$. Therefore, in accordance with Theorem 2.6,

$$u_p = \overline{\text{co}} \left(\bigcup_{t \in E_0} \{\mu_t\} \right).$$

Consider the set

$$S^+(E_0) \subset V(E):$$

$$S^+(E_0) = \{\mu \in V(E) \mid \mu \geqslant 0,\ \mu(E_0) = 1,\ \mu(E \setminus E_0) = 0\}.$$

One can easily show that $S^+(E_0)$ is convex. Let us show that $S^+(E_0)$ is weakly sequentially closed. We note first of all that $\mu \in S^+(E_0)$ if and only if the following three conditions are all satisfied: a) the inequality

$$\int_E x d\mu \geqslant 0$$

holds whenever x is an element of $C(E)$ such that $x(t) \geqslant 0$ for all $(t \in E)$: b) the equation

$$\int_E x d\mu = 0$$

holds whenever x is an element of $C(E)$ such that $x(t) = 0$ for all $(t \in E_0)$; c) for the element $\mathbf{1} \in C(E)$ defined for all $t \in E$ by $\mathbf{1}(t) = 1$

$$\int_E \mathbf{1} d\mu = 1.$$

Suppose now that $\mu_n \in S^+(E_0)$, and $\mu_n \xrightarrow{w} \mu$, meaning that, for every $x \in C(E)$:

$$\int_E x d\mu_n \longrightarrow \int_E x d\mu.$$

Then, if $x(t) \geqslant 0$ for every $(t \in E)$ we have

$$\int_E x d\mu_n \geqslant 0,$$

and hence

$$\int_E x d\mu \geqslant 0.$$

If $x(t) = 0$ for every $t \in E_0$;, then

$$\int_E x d\mu = 0;$$

and hence

$$\int_E x d\mu_n = 0.$$

Also,

$$\int_E 1 \, d\mu_n = 1,$$

and, therefore,

$$\int_E 1 \, d\mu = 1.$$

This proves the weak sequential closedness $S^+(E_0)$, and, consequently, the weak compactness of $S^+(E_0)$ since this set is bounded. Since $\mu_t \in S^+(E_0)$, where every $(t \in E_0)$, we have

$$u_p = \overline{\mathrm{co}}\left(\bigcup_{t \in E_0} \{\mu_t\}\right) \subset S^+(E_0).$$

On the other hand, if $\mu \in S^+(E_0)$, then, for every $x \in C(E)$,

$$\int_E x d\mu = \int_{E_0} x d\mu \leqslant \left(\max_{t \in E_0} x(t)\right) \mu(E_0) = \max_{t \in E_0} x(t) = p(x),$$

from which it follows that $\mu \in \mathfrak{u}_p$. Thus, we have shown that

$$\mathfrak{u}_p = \{\mu \in V(E) \mid \mu \geqslant 0, \ \mu(E_0) = 1, \ \mu(E \setminus E_0) = 0\}. \tag{2.12}$$

In particular, if $E_0 = E$, that is, if

$$p(x) = \max_{t \in E} x(t),$$

then

$$\mathfrak{u}_p = \{\mu \in V(E) \mid \mu \geqslant 0, \ \mu(E) = 1\}. \tag{2.13}$$

4) Let X denote a normed space, and let $f_1, f_2, \ldots, f_n$ denote members of X*. Let

$$p(x) = \sum_{k=1}^{n} |f_k(x)|.$$

Define $p_k(x) = |f_k(x)|$ $(k = 1, 2, \ldots, n)$. Since

$$p_k(x) = \max(f_k(x), -f_k(x)),$$

by using Theorem 2.6 we obtain

$$\mathfrak{u}_{p_k} = \{g \mid g = \alpha f_k, \ |\alpha| \leqslant 1\}$$

where $\mathfrak{u}_{p_k}$ is the smallest convex weakly compact set containing f_k and $-f_k$. It follows from Theorem 2.5 that

$$\mathfrak{u}_p = \mathfrak{u}_{p_1} + \mathfrak{u}_{p_2} + \ldots + \mathfrak{u}_{p_k} = \left\{ g \mid g = \sum_{k=1}^{n} \alpha_k f_k; \ |\alpha_k| \leqslant 1, \ k = 1, \ldots n \right\}.$$

A detailed description of the set of supporting linear functionals for a given sublinear functional is given in [25].

5°. Let p denote a sublinear functional defined on a normed space X. For $x \in X$, let

$$\Omega_x^p = \{y \in X \mid p(y) < p(x)\}.$$

In particular,

$$\Omega_0^p = \{y \in X \mid p(y) < 0\}.$$

One can easily show that if the set $\Omega_0^p \neq \varnothing$, then it is an open convex cone. Reference [25] presents the following interesting theorem, which describes the cone dual to Ω_0^p (see also [55]):

THEOREM 2.7. *If* $\Omega_0^p \neq \varnothing$ *then*[1] $(\Omega_0^p)^* = -C(u_p)$.

Suppose now that x is a member of X such that $\Omega_x^p \neq \varnothing$ (in other words, if p is a nonnegative functional, then $p(x) > 0$).

Let us define

$$G_x = \left\{g \in X^* \,\middle|\, \sup_{y \in \Omega_x^p} g(y) = g(x)\right\}.$$

THEOREM 2.8. *If* $\Omega_x^p \neq \varnothing$, *then*[2] $G_x = C(u_p^x)$.

Proof: 1) Suppose that $f \in C(u_p^x)$. Then, $f = \alpha g$, where $g \in u_p^x$, and $\alpha \geqslant 0$, Therefore, for $y \in \Omega_x^p$,

$$f(x) = \alpha g(x) = \alpha p(x) \geqslant \alpha p(y) \geqslant \alpha g(y) = f(y).$$

It follows that

$$f(x) \geqslant \sup_{y \in \Omega_x^p} f(y).$$

On the other hand, one can easily show that $x \in \bar{\Omega}_x^p$ (where $\bar{\Omega}_x^p$ is the closure of Ω_x^p). Therefore,

$$\sup_{y \in \Omega_x^p} f(y) = \sup_{y \in \bar{\Omega}_x^p} f(y) \geqslant f(x).$$

This means that $f \in G_x$.

[1]We recall that the set

$$(\Omega_0^p)^* = \{f \in X^* \mid f(x) \geqslant 0,\ x \in \Omega_0^p\}$$

is the cone dual to Ω_0^p and that $C(u_p) = \{\alpha f \mid \alpha \geqslant 0,\ f \in u_p\}$ is the conical hull of u_p.

[2]We recall that $u_p^x = \{f \in u_p \mid f(x) = p(x)\}$.

2) Suppose that $g \in G_x$ and $g \neq 0$. Let us show that

$$\operatorname{sign} g(x) = \operatorname{sign} p(x).$$

We consider three cases separately:

a) $p(x) > 0$. Suppose that $0 < \varepsilon < \frac{p(x)}{\|p\|}$. Then,[1] for $v \in S_\varepsilon(0)$,

$$p(v) \leqslant \|p\| \cdot \|v\| < \|p\| \cdot \frac{p(x)}{\|p\|} = p(x),$$

from which it follows that $S_\varepsilon(0) \subset \Omega_x^p$. We have

$$g(x) = \sup_{y \in \Omega_x^p} g(y) \geqslant \sup_{y \in S_\varepsilon(0)} g(y) > 0.$$

b) $p(x) < 0$. Suppose that $y \in \Omega_x^p$. Then, $p(y) < p(x)$. Therefore, for $\alpha \geqslant 1$, we have $p(\alpha y) = \alpha p(y) < p(x)$, from which it follows that $\alpha y \in \Omega_x^p$.

For $\alpha > 1$, the element αx is an interior point of the set Ω_x^p. This is true because, if $0 < \varepsilon < \frac{(1-\alpha) p(x)}{\|p\|}$ and $v \in S_\varepsilon(\alpha x)$, we obtain by using (2.4)

$$|p(v) - p(\alpha x)| \leqslant \|p\| \cdot \|v - \alpha x\| < (1-\alpha) p(x),$$

so that

$$p(v) < p(\alpha x) + (1-\alpha) p(x) = p(x).$$

If $g(x) \geqslant 0$,, we have $g(\alpha x) \geqslant 0$ when $\alpha > 1$. Therefore, as one can easily see,

$$\sup_{v \in S_\varepsilon(\alpha x)} g(v) > 0.$$

(Here, just as above, $0 < \varepsilon < \frac{(1-\alpha) p(x)}{\|p\|}$.)

Let v_0 denote a member of $\in S_\varepsilon(\alpha x)$ such that $g(v_0) > 0$. Then, since $\alpha v_0 \in \Omega_x^p$, for $\alpha \geqslant 1$, we have

[1] We recall that $S_\varepsilon(x)$ denotes the sphere of radius ε with center at the point x:

$$S_\varepsilon(x) = \{v \in X \mid \|v - x\| < \varepsilon\}.$$

$$\sup_{y \in \Omega_x^p} g(y) \geqslant \sup_{\alpha \geqslant 1} \alpha g(v_0) = \infty,$$

which is impossible. Thus, in the present case, $g(x) < 0$.

c) $p(x) = 0$. On the basis of our assumption,

$$\Omega_x^p \neq \varnothing.$$

It is clear that, in the present case, Ω_x^p coincides with the cone Ω_0^p mentioned in Theorem 2.7. Let us show that $g(x) = 0$. Since $0 \in \overline{\Omega}_x^p$, we have

$$g(x) = \sup_{y \in \Omega_x^p} g(y) = \sup_{y \in \overline{\Omega}_x^p} g(y) \geqslant 0.$$

On the other hand, if the inequality $g(y_0) > 0$ for some $y_0 \in \Omega_x^p$, then, since Ω_x^p is a cone,

$$g(x) = \sup_{y \in \Omega_x^p} g(y) \geqslant \sup_{\alpha \geqslant 0} \alpha g(y_0) = \infty,$$

which is impossible.

We now proceed to prove that $g \in C(\mathfrak{u}_p^x)$. Let us consider the case when $p(x) \neq 0$. Then, as was shown above, $g(x) \neq 0$ and sign $g(x) = \operatorname{sign} p(x)$.

Let us set $\lambda = \frac{p(x)}{g(x)}$. Obviously, $\lambda > 0$. The functional λg belongs to G_x and, furthermore, $\lambda g(x) = p(x)$. Let us show that λg belongs to $\mathfrak{u}_p$. Suppose that this is not the case. Then, there exists a $y \in X$ such that $\lambda g(y) > p(y)$. Without loss of generality, we may assume that

$$p(x) = \operatorname{sign} p(y). \tag{2.14}$$

If this is not the case, we can consider not y but the element $y_\alpha = x + \alpha(y - x)$. For this element,

$$p(y_\alpha) \leqslant \alpha p(y) + (1 - \alpha) p(x) < \lambda g(y_\alpha)$$

and, for sufficiently small α, we have $\operatorname{sign} p(y_\alpha) = \operatorname{sign} p(x)$ by virtue of the continuity of p. Assuming that (2.14) is satisfied, we can easily choose $\beta > 0$ such that

$$\lambda g(y) > \beta p(x) > p(y). \tag{2.15}$$

Now, the quantity $\left(\frac{1}{\beta}\right)y$ is a member of Ω_x^p. Therefore, since λg is a member of G_x, we have

$$\lambda g(x) = p(x) = \sup_{z \in \Omega_x^p} \lambda g(z) \geqslant \frac{1}{\beta} \lambda g(y),$$

which contradicts (2.15).

To complete the proof, we still need to consider the case when $p(x) = 0$. Then,

$$\sup_{y \in \Omega_x^p} g(y) = \sup_{y \in \Omega_0^p} g(y) = g(x) = 0,$$

from which it follows that g is a member of $-(\Omega_0^p)^*$.

It follows from Theorem 2.7 that $g \in C(u_p)$. Since, in addition, $\lambda g(x) = p(x) = 0$ for arbitrary $\lambda > 0$, there exists a $\lambda_0 > 0$ such that $\lambda_0 g \in U_p^x$. This completes the proof of the theorem.

6°. We shall say that a functional q is *superadditive* if

$$q(x_1 + x_2) \geqslant q(x_1) + q(x_2).$$

A superadditive positively homogeneous continuous functional is said to be *superlinear.* Obviously, if q is superlinear, the functional $p = -q$ is sublinear and, conversely, sublinearity of p implies superlinearity of $q = -p$. In view of this, we can easily carry over to superlinear functionals the results obtained in subsection 1° for sublinear functionals.

A functional $f \in X^*$ is said to be a supporting functional for a superlinear functional q if, for all $x \in X$,

$$f(x) \geqslant q(x). \tag{2.16}$$

The set of all functionals that are supporting for q will be denoted by u_q. If $p = -q$, then, as follows from (2.16), $u_q = -u_p$.

Suppose, as before, that q is superlinear. Let p' denote the functional defined by

$$p'(x) = -q(-x). \tag{2.17}$$

Obviously, p' is a sublinear functional and $\boldsymbol{u}^*_{p'} = \boldsymbol{u}_q$ [we note that, in turn, $q = -p'(-x)$].

Let us give some examples of superlinear functionals.

1) Suppose that $X = C(E)$, where E is a closed bounded set in n-dimensional space, and define

$$q(x) = \min_{t \in E} x(t).$$

We note that q is superlinear. Let

$$p'(x) = \max_{t \in E} x(t).$$

One can easily show that the functionals q and p' are related by equation (2.17).

2) Suppose that $X = L^\infty(E)$ where E is a subset of n-dimensional space. Let

$$q(x) = \operatorname{vrai}\min_{t \in E} x(t)$$

This q is a superlinear functional;

$$p'(x) = -q(-x) = \operatorname{vrai}\max_{t \in E} x(t).$$

We mention in conclusion that, if q is a superlinear functional, then, for $x \in X$,

$$q(x) = \min_{f \in \boldsymbol{u}_q} f(x). \tag{2.18}$$

To see this, if we set $p = -q$ and take (2.17) into account, we have

$$q(x) = -p(x) = -\max_{q \in \boldsymbol{u}_p} q(x) = \min_{q \in \boldsymbol{u}_p} (-q)(x) = \min_{f \in \boldsymbol{u}_p} f(x).$$

3. Differentiation of functionals

1°. Suppose that a functional f is defined on an open subset ξ of the space X. The derivative of the functional f at a point $x \in \xi$, evaluated on an element $u \in X$, is defined as the quantity

$$f'_x(u) = \lim_{\alpha \to +0} \frac{f(x+\alpha u) - f(x)}{\alpha}. \tag{3.1}$$

The existence of $f'_x(u)$ implies the existence of $f'_x(\lambda u)$ for $\lambda \geqslant 0$. Also, $f'_x(\lambda u) = \lambda f'_x(u)$.

The quantity

$$\frac{1}{\|u\|} f'_x(u) = f'_x\left(\frac{u}{\|u\|}\right)$$

is called the derivative of the functional f at the point x in the direction u. We shall say that a functional f is differentiable at a point x with respect to directions if the limit (3.1) exists for every $u \in X$. Differentiability of f with respect to directions at a point x means that there exists a functional f'_x defined on the space X such that, for arbitrary $u \in X$ and sufficiently small $\alpha > 0$,

$$f(x+\alpha u) = f(x) + \alpha f'_x(u) + o_{x,u}(\alpha), \tag{3.2}$$

where

$$\lim_{\alpha \to +0} \frac{o_{x,u}(\alpha)}{\alpha} = 0.$$

We shall call the functional f'_x the derivative of the functional f with respect to directions at the point x and we shall call (3.2) the formula on finite increments. Obviously, f'_x is a positively homogeneous functional.

We shall give yet another definition that will be important in what follows. Let f denote a functional that is differentiable with respect to directions at a point x. We shall say that f is uniformly differentiable with respect to directions at that point if the functions $o_{x,u}(\alpha)$ in equation (3.2) satisfy the following condition: For arbitrary $u \in X$ and arbitrary $\varepsilon > 0$, there exist $\delta_u > 0$ and $\alpha_u > 0$ such that the inequality

$$\left|\frac{o_{x,v}(\alpha)}{\alpha}\right|<\varepsilon$$

is satisfied for all $v\in S_{\delta_u}(u)$ and $\alpha\in(0,\ \alpha_u)$. Here,

$$S_{\delta_u}(u)=\{v\in X\,|\,\|v-u\|\leqslant\delta_u\}.$$

2°. Let us now look at the question of differentiability of a convex functional.

THEOREM 3.1. *A convex functional f defined on a normed space X is differentiable with respect to directions at every point x in X. Also, f'_x is a positively homogeneous semiadditive functional and the functions $o_{x,v}(\alpha)$ are nonnegative for every x and v in X.*

Proof: Suppose that x and u are members of X. Let α denote a real number. Consider the function $g_{x,u}(\alpha)=f(x+\alpha u)$. Let us show that this function is convex. For $t\in[0,\ 1]$,

$$\begin{aligned}g_{x,u}(t\alpha+(1-t)\beta)&=f(x+(t\alpha+(1-t)\beta)u)=\\&=f(t(x+\alpha u)+(1-t)(x+\beta u))\leqslant tf(x+\alpha u)+\\&+(1-t)f(x+\beta u)=tg_{x,u}(\alpha)+(1-t)g_{x,u}(\beta).\end{aligned}$$

As we know, a convex function defined on an interval has a finite right-hand derivative at every interior point of that interval (see [4, Chapter 1, Sec. 4]). It follows, in particular, that there exists the limit

$$\lim_{\alpha\to+0}\frac{g(\alpha)-g(0)}{\alpha}=\lim_{\alpha\to+0}\frac{f(x+\alpha u)-f(x)}{\alpha}=f'_x(u).$$

Thus, we have shown that the functional f has a derivative with respect to directions at every point x in X. We note that, as was shown in subsection 1°, the derivative with respect to directions is always a positively homogeneous functional.

Let us show that convexity of f implies its semiadditivity. We have

$$f'_x(u_1+u_2)=\lim_{\alpha\to+0}\frac{f(x+\alpha(u_1+u_2))-f(x)}{\alpha}=$$

$$=\lim_{\alpha\to+0}\frac{f\left(x+\frac{\alpha}{2}(u_1+u_2)\right)-f(x)}{\frac{\alpha}{2}}=\lim_{\alpha\to+0}\frac{2f\left(\frac{x+\alpha u_1}{2}+\frac{x+\alpha u_2}{2}\right)-2f(x)}{\alpha}\leqslant$$

$$\leqslant\lim_{\alpha\to+0}\frac{f(x+\alpha u_1)-f(x)}{\alpha}+\lim_{\alpha\to+0}\frac{f(x+\alpha u_2)-f(x)}{\alpha}=f'_x(u_1)+f'_x(u_2).$$

To complete the proof of the theorem, we still need to show that $o_{x,v}(\alpha) \geqslant 0$ for all x and v in X and all $\alpha > 0$. It follows from (3.2) that

$$\frac{o_{x,v}(\alpha)}{\alpha} = \frac{g_{x,v}(\alpha) - g_{x,v}(0)}{\alpha} - (g_{x,v})'_R(0), \tag{3.4}$$

where $g_{x,v}(\alpha) = f(x + \alpha v)$ and $(g_{x,v})'_R(0)$ denotes the right-hand derivative of $g_{x,v}$ at 0. Since $g_{x,v}$ is a convex function, the difference on the right side of (3.4) is nonnegative [see 4, Chapter 1, Sec. 4]. Our assertion then follows. This completes the proof of the theorem.

THEOREM 3.2.[1] *Let f denote a convex functional such that, for every $c > 0$,*

$$\sup_{\|x_1\| \leqslant c,\ \|x_2\| \leqslant c} \frac{|f(x_1) - f(x_2)|}{\|x_1 - x_2\|} = S(c) < \infty. \tag{3.5}$$

Then, the functional f is uniformly differentiable with respect to directions at an arbitrary point x and its derivative f'_x is a sublinear functional.

Proof: Let us show first that f'_x is a sublinear functional. It follows from Theorem 3.1 that f'_x is positively homogeneous and semiadditive. Therefore, we only need to show that it is bounded.

Let u denote a member of X such that $\|u\| = 1$. It follows from the formula on finite increments that, for $\alpha > 0$,

$$|f_x(u)| \leqslant \left| \frac{f(x + \alpha u) - f(x)}{\alpha} \right| + \left| \frac{o_{x,u}(\alpha)}{\alpha} \right|. \tag{3.6}$$

Since $\alpha = \|x + \alpha u - x\|$, it follows on the basis of (3.5) that

$$c_\alpha = \max(\|x + \alpha u\|,$$

where

$$\|x\|): \quad \left| \frac{f(x + \alpha u) - f(x)}{\alpha} \right| \leqslant S(c_\alpha).$$

Let us take the limit in (3.6), remembering that $c_\alpha < c = \|x\| + 1$ for $\alpha \in [0, 1]$ and that S is a nondecreasing function of c. We have $|f'_x(u)| \leqslant S(c)$, from which it follows that

[1]This theorem is given without proof in [25].

$$\sup_{\|u\|=1} |f'_x(u)| \leqslant S(c) < \infty.$$

This last inequality means that f'_x is a bounded functional. Let us show now that the functional f is uniformly differentiable. Suppose that x, u, and v are members of X. By using the formula on finite increments, we obtain

$$|o_{x,v}(\alpha) - o_{x,u}(\alpha)| = |f(x+\alpha v) - f(x) - \alpha f'_x(v) - f(x+\alpha u) + f(x) + \\ + \alpha f'_x(u)| \leqslant |f(x+\alpha v) - f(x+\alpha u)| + \alpha |f'_x(u) - f'_x(v)|.$$

It follows from (3.5) that

$$|f(x+\alpha v) + f(x+\alpha u)| \leqslant S(c_\alpha) \cdot \alpha \|u - v\|,$$

where $c_\alpha = \max(\|x+\alpha v\|, \|x+\alpha u\|)$. By using formula (2.4), we get

$$|f'_x(u) - f'_x(v)| \leqslant \|f'_x\| \cdot \|u - v\|.$$

Thus,

$$|o_{x,v}(\alpha) - o_{x,u}(\alpha)| \leqslant (S(c_\alpha) + \|f'_x\|) \cdot \alpha \|u - v\|. \tag{3.7}$$

Let ε denote a positive number. Let α_0 denote a number $\leqslant 1$ such that, for $0 < \alpha \leqslant \alpha_0$

$$\left| \frac{o_{x,u}(\alpha)}{\alpha} \right| \leqslant \frac{\varepsilon}{2}.$$

Define

$$\delta = \min\left(\frac{\varepsilon}{2(S(c) + \|f'_x\|)},\ 1 \right),$$

where $c = \|u\| + \|x\| + 1$. If v is such that $\|u - v\| < \delta$, then, *a fortiori* $\|u - v\| < 1$, so that $\|v\| \leqslant \|u\| + 1$. For these v and $\alpha \leqslant 1$,

$$c_\alpha = \max(\|x + \alpha v\|, \|x + \alpha u\|) \leqslant \max(\|x\| + \|v\|; \|x\| + \|u\|) \leqslant \\ \leqslant \|x\| + \|u\| + 1 = c,$$

and hence $S(c_\alpha) \leqslant S(c)$. In view of (3.7), we now have, for $v \in S_\delta(u)$ and $\alpha \leqslant \alpha_0$,

$$\left| \frac{o_{x,v}(\alpha)}{\alpha} \right| \leqslant \left| \frac{o_{x,v}(\alpha)}{\alpha} - \frac{o_{x,u}(\alpha)}{\alpha} \right| + \left| \frac{o_{x,u}(\alpha)}{\alpha} \right| \leqslant \\ \leqslant (S(c_\alpha) + \|f'_x\|) \cdot \|u - v\| + \frac{\varepsilon}{2} \leqslant (S(c) + \|f'_x\|)\delta + \frac{\varepsilon}{2} < \varepsilon,$$

which completes the proof.

3°. Let p denote a sublinear functional. It follows from formula (2.4) that p satisfies condition (3.5). (One can easily see that sub-linearity of p implies its convexity.) It follows from Theorem 3.2 that the functional p is uniformly differentiable at an arbitrary point x and that its derivative p'_x is a sublinear functional.

THEOREM 3.3. *A functional $f \in X$ is a supporting functional for p'_x if and only if $f \in u_p$ and $f(x) = p(x)$.*

Proof: 1) Suppose that $f \in u_p$ and $f(x) = p(x)$. Let us show that $f \in u_{p'_x}$. For $u \in X$,

$$f(x+\alpha u) \leqslant p(x+\alpha u) = p(x) + \alpha p'_x(u) + o_{x,u}(\alpha).$$

On the other hand,

$$f(x+\alpha u) = f(x) + \alpha f(u) = p(x) + \alpha f(u).$$

Thus,

$$p(x) + \alpha f(u) \leqslant p(x) + \alpha p'_x(u) + o_{x,u}(\alpha),$$

so that

$$f(u) \leqslant p'_x(u) + \frac{o_{x,u}(\alpha)}{\alpha}.$$

Taking the limit as $\alpha \to +0$, we see that $f \in U_{p'_x}$.

2) Suppose now that $f \in U_{p'_x}$. Then,

$$\alpha f(u) + p(x) + o_{x,u}(\alpha) \leqslant \alpha p'_x(u) + p(x) + o_{x,u}(\alpha) =$$
$$= p(x+\alpha u) \leqslant p(x) + \alpha p(u),$$

from which we get

$$\alpha f(u) + o_{x,u}(\alpha) \leqslant \alpha p(u).$$

If we divide this last inequality by α and take the limit as $\alpha \to +0$, we see that $f \in \mathfrak{u}_v$.

Let us show that $f(x) = p(x)$. For $\alpha \in [0, 1]$, we have

$$p(x - \alpha x) = p((1-\alpha)x) = (1-\alpha)p(x) = p(x) - \alpha p(x),$$

from which we get

$$p'_x(-x) = \lim_{\alpha \to +0} \frac{p(x-\alpha x) - p(x)}{\alpha} = -p(x).$$

Let us suppose now that $f(x) < p(x)$. Then,

$$f(-x) = -f(x) > -p(x) = p'_x(-x),$$

which contradicts the assumption that $f \in \mathfrak{u}_{p'_x}$. This completes the proof of the theorem.

Remark 1. We recall that we denoted by $\mathfrak{u}_p^x$ the set

$$\{f \in \mathfrak{u}_p \mid f(x) = p(x)\}.$$

Thus, Theorem 3.3 shows that $\mathfrak{u}_{p'_x} = \mathfrak{u}_p^x$. It follows from (2.7) that

$$p'_x(u) = \max_{f \in \mathfrak{u}_p^x} f(u).$$

From this it follows that, to define the derivative p'_x of the functional p at the point x, we need only know the set $\mathfrak{u}_p^x$.

Remark 2. Suppose that $p(x) = \|x\|$. Let us show that, in this case

$$\mathfrak{u}_p^x = \{f \in X^* \mid \|f\| = 1,\ f(x) = \|x\|\}.$$

Since $\mathfrak{u}_p = \{f \in X^* \mid \|f\| \leqslant 1\}$ [see (2.10)] in the present case, it follows on the basis of Theorem 3.3 that the set

$$\{f \in X^* \mid \|f\| = 1,\ f(x) = \|x\|\}$$

is contained in $\mathfrak{u}_p^x$.

Suppose now that $f \in u_p^x$. Then, $\|x\| = f(x) \leqslant \|f\| \cdot \|x\|$, from which it follows that $\|f\| > 1$. On the other hand, since $f \in u_p$, we have $\|f\| \leqslant 1$. Thus, $\|f\| = 1$, as we wished to show.

Remark 3. Suppose that p is a seminorm defined on a space X. Reasoning as in the proof of Remark 2 and using formula (2.11), one can easily show that

$$u_p^x = \{ f \in X^* \mid \|f\|_p = 1,\ f(x) = p(x) \}.$$

Here,

$$\|f\|_p = \sup_{p(x) \leqslant 1} |f(x)|.)$$

We note that the inequality $|f(x)| \leqslant \|f\|_p \, p(x)$ holds for arbitrary x in X.

4°. Let us evaluate the derivatives of certain sublinear functionals.

1) Suppose that $X = C(E)$ where E is, as usual, a closed bounded subset of R^n, and suppose that

$$p(x) = \|x\| = \max_{t \in E} |x(t)|.$$

Let x denote a member of $C(E)$ and let us define

$$E_1(x) = \{ t \in E \mid x(t) = p(x) \},$$
$$E_2(x) = \{ t \in E \mid x(t) = -p(x) \}, \quad E_3(x) = E \setminus (E_1(x) \cup E_2(x)).$$

Let us show that, in the present case, the set u_p^x consists of functionals generated by measures with the following properties:

$$\mu \geqslant 0 \text{ on the set } E_1(x), \tag{3.9}$$
$$\mu \leqslant 0 \text{ on the set } E_2(x), \tag{3.10}$$
$$\mu = 0 \text{ on the set } E_3(x), \tag{3.11}$$
$$\mu(E_1(x)) - \mu(E_2(x)) = 1. \tag{3.12}$$

Let us suppose to begin with that the functional f is generated by a measure possessing properties (3.9) - (3.12). Then,

$$\|f\| = \operatorname{var} \mu(E) = \mu(E_1(x)) - \mu(E_2(x)) = 1,$$
$$f(x) = \int_E x\,d\mu = \int_{E_1(x)} x\,d\mu + \int_{E_2(x)} x\,d\mu + \int_{E_3} x\,d\mu =$$
$$= p(x)\,\mu(E_1(x)) - p(x)\,\mu(E_2(x)) = p(x) = \|x\|.$$

It follows from Remark 2 following Theorem 3.3 that $f \in \mathfrak{u}_p^x$.

Suppose now that the functional f generated by the measure μ belongs to the set $\mathfrak{u}_p^x$. Using the same remark, we obtain the result

$$\begin{gathered} p(x)=f(x)=\int_E x d\mu=\int_{E_1(x)} x d\mu+\int_{E_2(x)} x d\mu+\int_{E_3(x)} x d\mu \leqslant \\ \leqslant p(x)\,\mu(E_1(x))-p(x)\,\mu(E_2(x))+p(x)\operatorname{var}\mu(E_3(x))= \\ =p(x)\left(\mu(E_1(x))-\mu(E_2(x))+\operatorname{var}\mu(E_3(x))\right) \leqslant \\ \leqslant p(x)\left(\operatorname{var}\mu(E_1(x))+\operatorname{var}\mu(E_2(x))+\operatorname{var}\mu(E_3(x))\right)= \\ =p(x)\operatorname{var}\mu(E), \end{gathered} \tag{3.13}$$

from which we see that $1 \leqslant \operatorname{var}\mu(E)$. Since $\operatorname{var}\mu(E)=\|\mu\|=\|f\|=1$, equality holds everywhere in (3.13). The fact that we can replace the symbol $\leqslant$ on its first occurrence with = signifies that var $\mu(E_3(x))=0$ or, what amounts to the same thing, that $\mu=0$ on $E_3(x)$. (In the opposite case,

$$\int_{E_3(x)} x d\mu < p(x)\operatorname{var}\mu(E_3(x))$$

since $x(t)<p(x)$ on $E_3(x)$.) The fact that on its second occurrence the symbol $\leqslant$ can be replaced with = means that

$$\mu(E_1(u))=\operatorname{var}\mu(E_1(x))$$

and

$$\mu(E_2(x))=-\operatorname{var}\mu(E_2(x))$$

or, what amounts to the same thing, that $\mu \geqslant 0$ on $E_1(x)$ and $\mu \leqslant 0$ on $E_2(x)$. Finally, equality in (3.13) is possible only when

$$\operatorname{var}\mu(E)=\operatorname{var}\mu(E_1(x))+\operatorname{var}\mu(E_2(x))=1.$$

Thus, conditions (3.9) - (3.12) are all satisfied.

By using formula (3.8) and properties (3.9) - (3.12) of measures defining functionals in U_p^x, we can easily show that, for arbitrary $u \in X$,

$$p_x'(u)=\max\left(\max_{t \in E_1(x)} u(t),\ \max_{t \in E_2(x)} -u(t)\right). \tag{3.14}$$

2) $X = C(E)$, E_0 is a closed subset of E, and

$$p(x) = \max_{t \in E_0} |x(t)|.$$

In this case, by using Remark 3 following Theorem 3.3, we can easily show that the set U_p^x consists of functionals f generated by measures μ possessing properties (3.9) - (3.12), where

$$E_1(x) = \{t \in E_0 \mid x(t) = p(x)\},$$
$$E_2(x) = \{t \in E_0 \mid x(t) = -p(x)\}, \quad E_3(x) = E \setminus (E_1(x) \cup E_2(x)).$$

One can easily show that p'_x can be calculated in accordance with formula (3.14).

3) $X = C(E)$ and

$$p(x) = \max_{t \in E} x(t).$$

With (2.13) in mind, we can show that

$$U_p^x = \{\mu \in V(E) \mid \mu \geqslant 0,\ \mu(E) = 1,\ \mu(E \setminus E(x)) = 0\},$$

where $E(x) = \{t \in E \mid x(t) = p(x)\}$. Suppose first that $\mu \in U_p$ and that μ is concentrated on the set $E(x)$, that is, $\mu(E \setminus E(x)) = 0$). Then,

$$\int_E x d\mu = \int_{E(x)} x d\mu = p(x)\,\mu(E(x)) = p(x),$$

that is, $\mu \in U_p^x$. Suppose now that $\mu \in U_p^x$. Then, $\mu \in U_p$; that is, $\mu \geqslant 0$ and $\mu(E) = 1$. Also,

$$p(x) = \int_E x d\mu = \int_{E(x)} x d\mu + \int_{E \setminus E(x)} x d\mu \leqslant p(x)\,\mu(E(x)) + \\ + p(x)\,\mu(E \setminus E(x)) = p(x)(\mu(E(x)) + \mu(E \setminus E(x))), \quad (3.15)$$

from which we get

$$\mu(E) = \mu(E(x)) + \mu(E \setminus E(x)) \geqslant 1. \quad (3.16)$$

Since $\mu(E) = 1$, equality must hold in (3.16), which is possible when

$$\mu(E \setminus E(x)) = 0.$$

One can easily show that, for any $u \in X$,

$$\max_{\mu \in u_p^x} \int_E u(t)\, d\mu = \max_{t \in E(x)} u(t),$$

from which it follows that

$$p'_x(u) = \max_{t \in E(x)} u(t). \tag{3.17}$$

4) $X = L^1(E)$ where E is a bounded measurable subset of n-dimensional space and

$$p(x) = \|x\| = \int_E |x(t)|\, dt.$$

Let us define

$$E_1(x) = \{t \in E \mid |x(t)| > 0\}, \quad E_2(x) = \{t \in E \mid |x(t)| = 0\}.$$

One can easily show by direct verification that, in the present case,

$$u_p^x = \left\{ f \in L^\infty(E) \mid f(t) = \begin{cases} \operatorname{sign} x(t), & t \in E_1(x) \\ v(t), & t \in E_2(x) \end{cases} \right\},$$

where v is an arbitrary measurable function defined on $E_2(x)$ that does not exceed unity in absolute value.

5) Let E denote the same set as in the preceding example. Suppose that $X = L^r(E)$, where $(1 < r < \infty)$, and suppose that

$$p(x) = \|x\| = \left(\int_E |x(t)|^r\, dt \right)^{\frac{1}{r}}.$$

By using the condition under which equality holds in the Hölder inequality, one can easily show that, in this case, the set u_p^x (for $x \neq 0$) consists of a single element; specifically,

$$u_r^x = \left\{ \left(\frac{|x|}{\|x\|} \right)^{r-1} \operatorname{sign} x \right\}.$$

5°. Suppose that q is a superlinear functional. Then, $p = -q$ is sublinear and, therefore,

$$q'_x(u) = -p'_x(u) = -\max_{f \in u^x_p} f(u) = \min_{f \in u^p_x} -f(u) = \min_{q \in u^q_x} q(u),$$

where

$$u^q_x = \{f \in u^q \mid q(x) = f(x)\}.$$

Let us find, for example, the derivative of the functional

$$q(x) = \min_{t \in E} x(t),$$

defined on the space $C(E)$. Suppose that

$$p(x) = \max_{t \in E} x(t).$$

Then, $q(x) = -p(-x)$ and, for $u \in X$, we have

$$q'_x(u) = \lim_{\alpha \to +0} \frac{q(x+\alpha u) - q(x)}{\alpha} =$$
$$= -\lim_{\alpha \to +0} \frac{p(-x-\alpha u) - p(-x)}{\alpha} = -p'_{-x}(-u).$$

In accordance with formula (3.17),

$$-p'_{-x}(-u) = -\max_{t \in E(-x)} (-u(t)) = \min_{t \in E(-x)} u(t),$$

where

$$E(-x) = \{t \in E \mid -x(t) = p(-x)\},$$

or, what amounts to the same thing,

$$E(-x) = \left\{t \in E \mid x(t) = \min_{\tau \in E} x(\tau)\right\}. \tag{3.18}$$

Thus,

$$q'_x(u) = \min_{t \in E(-x)} u(t), \tag{3.19}$$

where $E(-x)$ is defined by formula (3.18).

6°. Of special interest is the case in which the derivative f'_x of a functional f with respect to directions at a point x is a linear functional. In this case, we shall say that f is differentiable (differentiable in the sense of Gateau) at the point x, and we shall call the derivative the gradient (Gateau's gradient) of the functional f evaluated at the point x and shall denote it by Fx. If, in addition, the functions $o_{x,u}(\alpha)$ in (3.2) possess, for elements u whose norms are equal to unity, the property that

$$\frac{o_{x,u}(\alpha)}{\alpha} \xrightarrow[\alpha \to +0]{} 0$$

uniformly with respect to x, we shall say that f is differentiable in the sense of Fréchet at the point x. Obviously, in this case, the functional f is uniformly differentiable at the point x.

If a functional f is differentiable at every point of a subset ξ of X (in which case we shall say that f is differentiable on ξ), we have defined a mapping that maps every point x in ξ into the functional Fx in X^*. In other words, we have an operator F that maps ξ, into X^* in the following way: if $x \in \xi$, then Fx is a linear functional the value of which at each element μ is given by the formula

$$Fx(u) = \lim_{\alpha \to 0} \frac{f(x + \alpha u) - f(x)}{\alpha}. \tag{3.20}$$

(One can easily see that, in this case,

$$\lim_{\alpha \to +0} \frac{f(x + \alpha u) - f(x)}{\alpha} = \lim_{\alpha \to 0} \frac{f(x + \alpha u) - f(x)}{\alpha},$$

and hence the limit in (3.20) as $\alpha \to 0$ exists.) We shall call the operator F the gradient of the functional f.

If a functional f is differentiable on a convex set ξ, Lagrange's formula holds for it: if x and $x + u$ belong to ξ and if α belongs to $[0, 1]$, then

$$f(x + \alpha u) = f(x) + \alpha F(x + \theta u)\, u \quad (\theta \in [0, \alpha]). \tag{3.21}$$

To prove formula (3.21), one need only consider the function $g_{x,u}(\alpha) = f(x + \alpha u)$. One can easily show by direct calculation that

$g_{x,u}$ is differentiable and that $g'_{x,u}(\alpha)=F(x+\alpha u)\,u$. We have

$$f(x+\alpha u)-f(x)=g_{x,u}(\alpha)-g_{x,u}(0)=\alpha g'_{x,u}(\theta)=$$
$$=\alpha F(x+\theta u)\,u \quad (0\leqslant\theta\leqslant\alpha),$$

which is what we wished to show.

If f is a convex functional, the function $g_{x,u}$ is convex, as one can easily show. Therefore, its derivative $g'_{x,u}$ is a nondecreasing function. From this it follows, in particular, that if f is a convex functional, we have, for arbitrary x, and $x+u$ in ξ and α in [0, 1],

$$f(x+\alpha u)\geqslant f(x)+\alpha Fx(u). \tag{3.22}$$

We present without proof the following important formula, which enables us to determine a functional from its gradient [6, Remark 6.1]. If x and $x+u$ belong to an open convex set ξ, then

$$f(x+u)=f(x)+\int_0^1 F(x+\alpha u)(u)\,d\alpha. \tag{3.23}$$

We note also the validity of the following important fact [6, Theorem 3.3]: if an operator F is the gradient of a functional f and if F is continuous on a set ξ, then the functional f is differentiable in the sense of Fréchet on ξ.

7°. Suppose that X and Y are normed spaces, that ξ is an open subset of X, and that A is an operator mapping ξ into Y.

We shall say that A is differentiable with respect to directions at a point $x\in\xi$ if there exists an operator A'_x that maps X and Y and, for $u\in X$ and sufficiently small $\alpha>0$, satisfies the equation

$$A(x+\alpha u)=Ax+\alpha A'_x u+o_{x,u}(\alpha), \tag{3.24}$$

where

$$\frac{o_{x,u}(\alpha)}{\alpha}\xrightarrow[\alpha\to+0]{}0.$$

It follows from formula (3.24) that

$$A'_x u=\lim_{\alpha\to+0}\frac{A(x+\alpha u)-Ax}{\alpha}.$$

We shall call the operator A'_x the derivative of the operator A with respect to directions at the point x.

Of especial interest is the case when A'_x is a linear operator. We shall say in this case that A is differentiable (differentiable in the sense of Gateau) at a point x and we shall call the operator A'_x the Gateau derivative. We note that, in the present case,

$$A'_x(u) = \lim_{\alpha \to 0} \frac{A(x + \alpha u) - Ax}{\alpha}.$$

If

$$\frac{\| o_{x, u}(\alpha) \|}{\alpha} \to 0$$

uniformly with respect to u, where u lies in a unit sphere, we say that A is differentiable in the sense of Fréchet. If the operator A is differentiable on some convex set η, then Lagrange's formula holds for it in the following "weak" form: if x and $x + u$ belong to η and α belongs to $[0, 1]$, then, for arbitrary $h \in X^*$,

$$h(A(x + \alpha u)) = h(Ax) + \alpha h(A'_{x+\theta_h u}(u)), \quad \theta_h \in [0, \alpha]. \qquad (3.25)$$

To prove (3.25), it will be sufficient to consider the functional $f(x) = h(Ax)$ and to apply Lagrange's formula [see (3.21)] to it. (One can easily show that f is differentiable on η and that $Fx(u) = h(A'_x(u)$ for every $u \in X$.)

Let us give some examples of differentiable operators.

1) Suppose that $X = R^n$, $Y = R^m$, ξ is an open subset of R^n, and E is a closed bounded subset of R^m. Let $g(x, y)$ denote a function that is continuous with respect to y on the set

$$\xi \times E = \{(x, y) \mid x \in \xi, \ y \in E\}.$$

Suppose that $g(x, y)$ has a partial derivative with respect to x

$$\frac{\partial g}{\partial x} = \left(\frac{\partial g}{\partial x_1}, \frac{\partial g}{\partial x_2}, \ldots, \frac{\partial g}{\partial x_n} \right)$$

that is continuous with respect to y at every point (x, y) for $x \in \xi$ and $y \in E$.

Let us define an operator A_1 mapping the set ξ into the space $C(E)$ as follows: if $x_0 \in \xi$, then $Ax_0 = z \in C(E)$, where $z(y) = g(x_0, y)$. Let us show that, if A is differentiable at an arbitrary point x_0 in ξ, then, for arbitrary $u = (u_1, u_2, \ldots, u_n) \in R^n$,

$$(A'_{x_0}(u))(y) = \frac{\partial g}{\partial x}(x_0, y)(u) = \sum_{i=1}^{n} \frac{\partial g}{\partial x_i}(x_0, y) u_i. \tag{3.26}$$

We have

$$(A'_{x_0}(u))(y) = \lim_{\alpha \to +0} \frac{(A(x_0 + \alpha u))(y) - (Ax_0)(y)}{\alpha} =$$
$$= \lim_{\alpha \to +0} \frac{g(x_0 + \alpha u, y) - g(x_0, y)}{\alpha} = \frac{\partial g}{\partial x}(x_0, y)(u).$$

The linearity of the operator A'_{x_0} is obvious.

2) Let E denote a closed bounded subset of R^n. Consider a function $g(x, t)$ defined for all real x and $t \in E$ that is continuous in its domain of definition and that has in that domain a continuous partial derivative $\frac{\partial g}{\partial x}$. Let us define an operator A mapping $C(E)$ into itself as follows: if $x(t) \in C(E)$, then $(Ax)(t) = g(x(t), t)$. (The fact that $Ax \in C(E)$ and also the continuity and boundedness of A are pointed out in [36, Chapter I, Theorem 2.3]; these follow only from the continuity of $g(x, t)$.) Let us show that if A is a differentiable operator, then

$$(A'_x(u))(t) = \frac{\partial g}{\partial x}(x(t), t) u(t). \tag{3.27}$$

We have

$$(A'_x(u))(t) = \lim_{\alpha \to +0} \frac{A(x + \alpha u)(t) - Ax(t)}{\alpha} =$$
$$= \lim_{\alpha \to 0} \frac{g(x(t) + \alpha u(t), t) - g(x(t), t)}{\alpha}.$$

If a point t_0 is such that $u(t_0) = 0$, then $A_x(u)(t_0) = 0$ as follows from the equation written above. On the other hand,

$$\frac{\partial g}{\partial x}(x(t_0), t_0) u(t_0) = 0.$$

If $u(t_0)=0$, then

$$(A'_x(u))(t_0)=\lim_{\alpha\to+0}\frac{g(x(t_0)+\alpha u(t_0),\ t_0)-g(x(t_0),\ t_0)}{\alpha u(t_0)}u(t_0)=$$
$$=\frac{\partial g}{\partial x}(x(t_0),\ t_0)\,u(t_0),$$

so that equation (3.27) is proven. The additivity and homogeneity of A'_x are obvious. That A'_x is bounded follows from the fact that the operator of multiplication by a fixed function is continuous in $C(E)$.

3) Suppose that E is a closed bounded subset of R^n. Let $g(x,\ y,\ t)$ denote a function defined for all real x and y and for $t\in E$. Suppose that this function is continuous and has continuous partial derivatives $\frac{\partial g}{\partial x}$ and $\frac{\partial g}{\partial y}$ in its domain of definition. Consider the space $X\times Y$ where X and Y are either $C(E)$ or $L^\infty(E)$. Let us define an operator A mapping $X\times Y$ into $L^\infty(E)$ as follows:

$$(A(x,\ y))(t)=g(x(t),\ y(t),\ t).$$

Reasoning as above, one can easily show that, if A is a differentiable operator, then

$$\begin{aligned}(A'_{(x,y)}(u,\ v))(t)=&\frac{\partial g}{\partial x}(x(t),\ y(t),\ t)\,u(t)+\\ &+\frac{\partial g}{\partial y}(x(t),\ y(t),\ t)\,v(t).\end{aligned}\tag{3.28}$$

8°. Let f denote a functional defined on a space X and differentiable in the sense of Gateau on an open subset ξ of X. Then, if the operator F (the gradient of f) is differentiable at a point $x\in\xi$, we shall say that f is twice differentiable at the point x, and we shall call F'_x the second derivative of f at the point x.

If ξ is convex and f is differentiable at every point ξ of x (in case we shall, as usual, say that f is twice differentiable on ξ), then the following formulas for the Taylor expansion are valid: For x and $x+u$ in ξ and α in $[0,\ 1]$,

$$f(x+\alpha u)=f(x)+\alpha Fx(u)+\frac{\alpha^2}{2}[F'_x(u)](u)+o(\alpha^2),\tag{3.29}$$

$$f(x+\alpha u)=f(x)+\alpha Fx(u)+\frac{\alpha^2}{2}[F'_{x+\theta u}(u)](u)\quad(\theta\in[0,\ \alpha]).\tag{3.30}$$

One can prove these formulas by expanding the function

$$g_{x,u}(\alpha)=f(x+\alpha u)$$

in a Taylor series, writing out the first two terms. (Under our assumptions, $g_{x,u}$ is a twice differentiable function and

$$g'_{x,u}(\alpha)=F(x+\alpha u)\,u, \quad g''_{x,u}(\alpha)=\left[F'_{x+\alpha u}(u)\right](u))$$

9°. Let us look at the question of the differentiability of a "composite" function. Let X and Y denote normed spaces, let A denote an operator mapping a subset ξ of X with a nonempty interior into the space Y, and let g denote a functional defined on Y.

Consider the functional f defined by $f(x)=g(Ax)$ on the set ξ. Let x denote a member of $\mathring{\xi}$. Then, if A is differentiable with respect to directions at the point x, if g is uniformly differentiable with respect to directions at the point Ax, and if the functional g'_{Ax} is continuous, then f is differentiable with respect to directions at x and, for $u \in X$, we have

$$f'_x(u)=g'_{Ax}(A'_x u). \tag{3.31}$$

Let us prove this assertion. We have

$$\frac{f(x+\alpha u)-f(x)}{\alpha}=\frac{g(A(x+\alpha u))-g(Ax)}{\alpha}=$$

$$=\frac{g(Ax+\alpha A'_x(u)+o_{x,u}(\alpha))-g(Ax)}{\alpha}=$$

$$=\frac{g\left(Ax+\alpha\left(A'_x(u)+\frac{o_{x,u}(\alpha)}{\alpha}\right)\right)-g(Ax)}{\alpha}=$$

$$=\frac{g(Ax)+\alpha g'_{Ax}\left(A'_x(u)+\frac{o_{x,u}(\alpha)}{\alpha}\right)+\tilde{o}_{Ax,v(\alpha)}(\alpha)-g(Ax)}{\alpha}=$$

$$=g'_{Ax}\left(A'_x(u)+\frac{o_{x,u}(\alpha)}{\alpha}\right)+\frac{\tilde{o}_{Ax,v(\alpha)}(\alpha)}{\alpha}, \tag{3.32}$$

where

$$v(\alpha)=A'_x(u)+\frac{o_{x,u}(\alpha)}{\alpha},$$

$$o_{Ax,v(\alpha)}(\alpha)=g(Ax+\alpha v)-g(Ax)-\alpha g'_{Ax}(v).$$

By virtue of the uniform differentiability of g, for every ε, there exist δ and α_0 such that, for all α in $(0, \alpha_0]$ and all v satisfying

$$\left|\frac{\tilde{o}_{Ax,\,v}(\alpha)}{\alpha}\right| < \varepsilon.$$

Remembering that

$$\frac{o_{x,\,u}(\alpha)}{\alpha} \underset{\alpha\to+0}{\to} 0,$$

we obtain

$$\left\| v(\alpha) - A'_x(u) \right\| = \left\| \frac{o_{x,\,u}(\alpha)}{\alpha} \right\| \leqslant \delta,$$

for sufficiently small positive α. Consequently, for these α,

$$\left|\frac{\tilde{o}_{Ax,\,v(\alpha)}(\alpha)}{\alpha}\right| < \varepsilon.$$

In view of this last inequality and the continuity of the functional g'_{Ax}, we obtain by taking the limit in (3.32),

$$f'_x(u) = \lim_{\alpha\to+0} \frac{f(x+\alpha u) - f(x)}{\alpha} = g'_{Ax}(A'_x(u)),$$

which is what we wished to show.

10°. Let us give some examples that will be important in what follows.

1) Let $g(x, y)$ denote the function considered in example 1) of subsection 7° of this section. Let f_1 and f_2 denote functionals defined on a subset ξ of R^n by the following equations: for $x \in \xi$

$$f_1(x) = \max_{y \in E} g(x, y), \quad f_2(x) = \min_{y \in E} g(x, y).$$

Assuming that A is differentiable and using formulas (3.26) and (3.31) and also (3.17) and (3.19), we have

$$(f_2)'_x(u) = \max_{y \in E_1(x)} \frac{\partial g}{\partial x}(x, y)(u), \tag{3.33}$$

where

$$E_1(x) = \{y \in E \mid g(x, y) = \max_{z \in E} g(x, z)\},$$

and

$$(f_2)'_x(u) = \min_{y \in E_2(x)} \frac{\partial g}{\partial x}(x, y)(u), \tag{3.34}$$

where

$$E_2(x) = \{y \in E \mid g(x, y) = \min_{z \in E} g(x, z)\}.$$

2) Let $g(x, t)$ denote the function considered in example 2) of subsection 7°. Let f_1 and f_2 denote the functionals defined on the space $G(E)$ according to the following formulas: for $x \in G(E)$,

$$f_1(x) = \max_t g(x(t), t), \quad f_2(x) = \min_t g(x(t), t).$$

Assuming that A is differentiable and using formulas (3.27) and (3.31) and also (3.17) and (3.19), we have

$$(f_1)'_x(u) = \max_{t \in E_1(x)} \frac{\partial g}{\partial x}(x(t), t)\, u(t), \tag{3.35}$$

where

$$E_1(x) = \{t \in E \mid g(x(t), t) = \max_{\tau \in E} g(x(\tau), \tau)\};$$

and

$$(f_2)'_x(u) = \min_{t \in E_2(x)} \frac{\partial g}{\partial x}(x(t), t)\, u(t), \tag{3.36}$$

where

$$E_2(x)=\{t\in E \mid g(x(t),t)=\min_{\tau\in E} g(x(\tau),\tau)\}.$$

3) Suppose that

$$f(x,y)=\int_E g(x(t),\ y(t),\ t)\,dt$$

is a functional defined on the space of pairs of functions $X\times Y$ where X and Y are either $G(E)$ or $L^\infty(E)$, by the function $g(x,\ y,\ t)$ considered in example 3) of subsection 7°. Remembering that the functional φ defined by

$$\varphi(u)=\int_E u\,dt$$

is linear in $L^\infty(E)$ and using formulas (3.28) and (3.31) and the remark following example 1) of subsection 4°, we have

$$f'_{x,y}(u,\ v)=\int_E\left[\frac{\partial g}{\partial x}(x(t),\ y(t),\ t)\,u(t)+\frac{\partial g}{\partial y}(x(t),\ y(t),\ t)\,v(t)\right]dt.$$

4. Conditions for existence and uniqueness of an extremum. Minimizing sequences

In this section, we shall consider conditions for existence and uniqueness of a minimum of a functional f defined on a normed space X. All the results that we shall obtain can be carried over to the case of a maximum with natural changes.

1°. The fundamental results regarding the existence of a minimum are given by the following theorem, which is a generalization of Weierstrass' theorem that every continuous function is bounded on a closed interval and attains its greatest lower bound.

THEOREM 4.1 (Weierstrass). *Let Ω denote a compact (resp. weakly compact) subset of a Banach space X and let f denote a lower-semicontinuous (resp. weakly lower-semicontinuous functional. Then f is bounded below on Ω and attains a minimum on Ω.*

Proof: For definiteness, let us assume that Ω is compact and that f is lower-semicontinuous. (If Ω is weakly compact and f is weakly lower-semicontinuous, the proof will remain valid with obvious modifications.) We shall also assume that Ω is an infinite set.

Let us suppose that

$$\inf_{x \in \Omega} f(x) = -\infty.$$

Then, there exists a sequence of elements x_n of Ω such that

$$\lim_{n \to \infty} f(x_n) = -\infty.$$

Since Ω is compact, this sequence contains a convergent subsequence of elements x_{n_k} the limit of which is an element of Ω. Since f is lower-semicontinuous, we have

$$f(x) \leqslant \underline{\lim}_{k \to \infty} f\left(x_{n_k}\right) = \lim_{n \to \infty} f(x_n) = -\infty,$$

which is impossible. Thus,

$$\inf_{x \in \Omega} f(x) = m > -\infty.$$

It follows from the definition of an infimum that there exists a sequence of elements y_n of Ω such that

$$m = \lim_{y_n \to \infty} f(y_n).$$

This sequence contains a convergent subsequence of elements y_{n_k}, the limit of which we denote by y. It follows from the lower-semicontinuity of f that

$$f(y) \leqslant \underline{\lim}\, f\left(y_{n_k}\right) = \lim f(y_n) = m.$$

On the other hand, since $y \in \Omega$, we have $f(y) \geqslant m$. Thus,

$$f(y) = m = \inf_{x \in \Omega} f(x),$$

which completes the proof.

COROLLARY. *Every quasiconvex (and a fortiori convex) functional f defined on a Banach space X attains a minimum on a weakly compact subset x of X. In particular, if X is a reflective space, f attains a minimum on an arbitrary closed convex bounded set* (see Theorem 1.1).

Remark 1. If f is a strictly quasiconvex (and *a fortiori* if f is a strictly convex) functional, it attains a minimum on a convex weakly compact set Ω at a unique point. This is true because if we had

$$\min_{x \in \Omega} f(x) = f(y_1) = f(y_2)$$

where y_1 and y_2 are distinct members of Ω, we would have

$$f\left(\frac{y_1 + y_2}{2}\right) < \max\{f(y_1), f(y_2)\} = \min_{x \in \Omega} f(x),$$

which is impossible.

Remark 2. There is a unique minimum also in the case in which Ω is a strictly convex weakly compact set and f is a convex functional possessing the property that, for every x in $\mathring{\Omega}$, there exists a u in X such that $f'_x(u) < 0$. To see this, let us suppose that

$$\min_{z \in \Omega} f(z) = f(y_1) = f(y_2)$$

where y_1 and y_2 are distinct members of Ω. It follows from the strict convexity of Ω that the point $x = \frac{y_1 + y_2}{2}$ is an interior point of Ω. Therefore, there exists a u in X such that $f'_x(u) < 0$. Let α denote a positive number sufficiently small that $x + \alpha u \in \Omega$ and

$$o_{x,u}(\alpha) < \frac{1}{2} \alpha \left| f'_x(u) \right|$$

(Here, $o_{x,u}(\alpha)$ is the function defined for the functional f by means of the formula on finite increments (3.2).) By using (3.2), we have

$$\frac{f(x + \alpha u) - f(x)}{\alpha} = f'_x(u) + \frac{o_{x,u}(\alpha)}{\alpha} < \frac{1}{2} f'_x(u) < 0,$$

from which it follows that

$$\min_{z \in \Omega} f(z) \leqslant f(x + \alpha u) < f(x) \leqslant \frac{1}{2}(f(y_1) + f(y_2)) = \min_{z \in \Omega} f(z),$$

which is impossible.

We note that, if f is a differentiable functional, our assumption regarding f is equivalent to the assumption that $\|Fx\| \neq 0$ for $x \in \overset{\circ}{\Omega}$ (here, as usual, Fx denotes the gradient of f at the point x).

2°. Let Ω denote a bounded subset of a Banach space X and suppose that f is defined on X. We shall call a sequence $\{x_n\}$, where $x_n \in \Omega$, $n = 1, 2, \ldots$, a *minimizing sequence* if

$$\lim_{n \to \infty} f(x_n) = \inf_{x \in \Omega} f(x).$$

In the proof of Theorem 4.1, it was also proven in effect that, if Ω is a compact (resp. weakly compact) set and f is lower-semicontinuous (resp. weakly lower-semicontinuous) on Ω, then an arbitrary minimizing sequence $\{x_n\}$ contains a subsequence $\{x_{n_k}\}$ that converges (resp. converges weakly) to the point at which minimum is attained. In particular, if the conditions of Remarks 1 and 2 following Theorem 4.1 are satisfied, then, as one can easily see, every minimizing sequence converges weakly to the (unique) point at which the minimum of f is attained on Ω.

Let us suppose that a quasiconvex functional f satisfies the following condition:

$$f\left(\frac{x+y}{2}\right) \leqslant \max\{f(x), f(y)\} - \delta(\|x-y\|), \tag{4.1}$$

where $\delta(\tau)$ is an increasing continuous function and $\delta(0) = 0$. (In [44], such functionals are called *uniformly quasiconvex functionals.*)

Obviously, a functional f satisfying conditions (4.1) is strictly quasiconvex and hence attains a minimum on a weakly compact convex set Ω at a unique point. We have the following theorem (see [44]):

THEOREM 4.2. *If a functional f satisfies condition (4.1) and attains a minimum on a convex weakly compact set Ω at a point y, then an arbitrary minimizing sequence converges (with respect to the norm) to Y.*

Proof: Let $\{x_n\}$ denote a minimizing sequence:

$$\lim f(x_n) = \min_{x \in \Omega} f(x) = f(y).$$

Then, by virtue of (4.1),

$$f\left(\frac{x_n+y}{2}\right) \leqslant \max(f(x_n), f(y)) - \delta(\|x_n - y\|) = \\ = f(x_n) - \delta(\|x_n - y\|), \tag{4.2}$$

so that

$$\delta(\|x_n - y\|) \leqslant f(x_n) - f\left(\frac{x_n+y}{2}\right). \tag{4.3}$$

From (4.2) it also follows that

$$f\left(\frac{x_n+y}{2}\right) < f(x_n).$$

Let us show that

$$\lim_{n\to\infty} \delta(\|x_n - y\|) = 0.$$

Let us suppose the contrary. Then, there exists an $\varepsilon > 0$ and a subsequence (x_{n_k}) such that

$$\delta\left(\left\|x_{n_k} - y\right\|\right) > \varepsilon.$$

We may assume without loss of generality that the sequence

$$f\left(\frac{x_{n_k}+y}{2}\right)$$

converges. Obviously, the limit of this sequence coincides with $f(y)$, so that

$$\lim\left(f\left(x_{n_k}\right) - f\left(\frac{x_{n_k}+y}{2}\right)\right) = 0.$$

This, however, contradicts (4.3). Thus,

$$\lim_{n\to\infty} \delta(\|x_n - y\|) = 0.$$

Then, by virtue of the properties of the function δ, it follows that $x_n \to y$. This completes the proof of the theorem.[1]

We conclude this section with yet another case in which convergence (with respect to a norm) of minimizing sequences is guaranteed. Let us suppose that a convex weakly compact set Ω possesses the property that an increasing function $\delta(\tau)$ with $\delta(0)=0$ exists such that

$$\frac{x+y}{2}+z \in \Omega, \tag{4.4}$$

whenever x and y belong to Ω and $\|z\| \leqslant \delta(\|x-y\|)$. (In [44], such sets are called *uniformly convex.*) Obviously, a set satisfying condition (4.4) is strictly convex. Remembering that a convex functional is differentiable with respect to directions at every point, let us prove the following theorem:

THEOREM 4.3.[2] *Suppose that a convex functional f attains a minimum on a set w satisfying condition (4.4) at a point y. Suppose that*

$$\inf_{\|u\| \leqslant 1} f_y'(u) = -\varepsilon < 0. \tag{4.5}$$

Then, for every $x \in \Omega$,

$$f(x)-f(y) \geqslant 2\varepsilon\delta(\|x-y\|). \tag{4.6}$$

Proof: Since the minimum is attained at the point y, it follows that

$$f(y+\alpha(x-y)) = f(y)+\alpha f_y'(x-y)+o_{y,\,x+y}(\alpha) \geqslant f(y),$$

for every x in Ω and every α in $[0, 1]$. If follows[3] that $f_y'(x-y) \geqslant 0$. Since Ω satisfies condition (4.4), we have

[1] An example of a functional satisfying condition (4.1) is the norm in the space $L^2(E)$.
[2] Compare with Theorem 1.6 in [44].
[3] We have thus found a necessary condition for a minimum of (2.2.11).

$$f'_y\left(\frac{x+y}{2}+\delta(\|x-y\|)u-y\right)\geqslant 0$$

for every x in Ω in $S_1(0)$, where $(S_1(0)=\{u\in X \mid \|u\|\leqslant 1\})$. Remembering that the functional f'_y is positively homogeneous and semiadditive (see Theorem 3.1), we obtain

$$\frac{1}{2}f'_y(x-y)+\delta(\|x-y\|)f'_y(u)\geqslant$$
$$\geqslant f'_y\left(\frac{x+y}{2}+\delta(\|x-y\|)\,u-y\right)\geqslant 0. \tag{4.7}$$

On the other hand,

$$f(x)=f(y+(x-y))=f(y)+f'_y(x-y)+o_{y,\,x-y}(1)\geqslant$$
$$\geqslant f(y)+f'_y(x-y). \tag{4.8}$$

(Here, we used the fact that, by virtue of the convexity of f, $o_{x,\,x+y}(1)\geqslant 0$.) By using (4.7) and (4.8), we obtain

$$f(x)-f(y)\geqslant f'_y(x-y)\geqslant -2\delta(\|x-y\|)f'_y(u).$$

Since this last inequality holds for all $u\in S_1(0)$, we have

$$f(x)-f(y)\geqslant 2\delta(\|x-y\|)\sup_{\|u\|\leqslant 1}(-f'_y(u))=$$
$$=2\delta(\|x-y\|)\left(-\inf_{\|u\|\leqslant 1}f'_y(u)\right)=2\varepsilon\delta(\|x-y\|),$$

which completes the proof.

COROLLARY. *Suppose that the conditions of Theorem 4.3 are satisfied and that* $\{x_n\}$ *is a minimizing sequence. Then, it follows from (4.6) and the properties of the function* δ *that* $\{x_n\}$ *converges, with respect to norm, to the point* y.

Remark. If f is a differentiable functional, condition (4.5) is equivalent to

$$\|Fy\|=\sup_{\|u\|\leqslant 1}Fy(u)=\sup_{\|u\|<1}Fy(-u)=$$
$$=\sup_{\|u\|\leqslant 1}(-Fy(u))=-\inf_{\|n\|<1}Fy(u)=\varepsilon>0.$$

Here, Fy is the gradient of f at the point y. The inequality $\|Fy\|>0$ is, as follows from the results of Chapter 2, a condition for the absolute minimum of f over all X not to be attained at the point y. We note that the unit sphere in $L^2(E)$ is an example of a set satisfying condition (4.4).

5. Minimization of a linear integral functional

1°. Let Ω denote a convex weakly compact subset of a Banach space X. Then, it follows from Theorem 4.1 that every linear functional h attains a minimum on Ω. In general, the problem of finding the points $x\in\Omega$ at which h attains its minimum on Ω is a very complicated one. However, for certain sets Ω that are important in applications, this problem can be solved in a comparatively simple way.

Suppose that $h\in X^*$. Define G_Ω by

$$G_\Omega h=\{x\in\Omega \mid h(x)=\min_{y\in\Omega} h(y)\}. \tag{5.1}$$

This G_Ω can be regarded as a mapping that maps a linear functional into a subset of Ω. Let us list some properties of this mapping.

1) $G_\Omega(0)=\Omega$.

2) If $\lambda>0$, then $G_\Omega(\lambda h)=G_\Omega h$.

3) Let $\Omega'=a+\lambda\Omega$ (where $a\in X$, $\lambda>0$). Then,

$$G_{\Omega'}=a+\lambda G_\Omega. \tag{5.2}$$

4) If $h_n\to h$ and $x_n\to x$, where $x_n\in G_\Omega h_n$, then $x\in G_\Omega h$ (here, one or the other of the sequences may converge weakly). To see this, note that since $x_n\in G_\Omega h_n$, we have $h_n(x_n)\leqslant h_n(y)$, where $(y\in\Omega)$. By taking the limit, we obtain the assertion made. It is convenient to call this property of the mapping G_Ω its weak upper-semicontinuity.

5) If Ω is a symmetric set (that is, if $-x$ belongs to it whenever x does), then $G_\Omega(-h)=-G_\Omega h$ for $h\in X^*$.

6) Suppose that X and Y are Banach spaces and that B is a linear operator mapping X into Y. Suppose that Ω' is a convex weakly compact subset of X and

$$\Omega=B(\Omega')\equiv\{y\in Y \mid y=Bx,\ x\in\Omega'\}.$$

Then, for $h \in Y^*$,

$$G_\Omega h = B(G_{\Omega'} B^* h) = \{y \in \Omega \mid y = Bx,\ x = G_{\Omega'} B^* h\}. \tag{5.3}$$

Here, B^* is the dual of the operator B.

We note first of all that the weak compactness of Ω' implies the weak compactness of Ω. Therefore, for any $h \in Y^*$, we have $G_\Omega h \neq \varnothing$. The validity of (5.3) follows from the following relationships for $x_0 \in G_{\Omega'} B^* h$:

$$\min_{y \in \Omega} h(y) = \min_{x \in \Omega'} h(Bx) = \min_{x \in \Omega'} B^* h(x) = B^* h(x_0) = h(Bx_0).$$

Since x_0 is an arbitrary element of the set $G_{\Omega'} B^* h$, we have

$$G_\Omega h = B(G_{\Omega'} B^* h).$$

We mention, that, in what follows, it will usually be sufficient for us to know some one element of the set $G_\Omega h$ and not the entire set (in other words, it will be sufficient for us to know one of the elements at which h attains its minimum on Ω).

2°. Let us describe the mapping G_Ω in the case in which Ω is a unit sphere in a reflexive space X. (As was noted in Sec. 1, in the present case Ω is weakly compact.) Let us consider X as the space dual to X^*. We denote by $\|\cdot\|^*$ the norm in the space X^*. Let us show that, for an arbitrary nonzero $h \in X^*$,

$$G_\Omega h = -U^h_{\|\cdot\|^*}. \tag{5.4}$$

If $y \in U^h_{\|\cdot\|^*}$, then (see Remark 2 following Theorem 3.3) we have $\|y\| = 1$ (that is, $y \in \Omega$), and, furthermore, $y(h) = \|h\|$. Therefore, we have

$$\min_{x \in \Omega} h(x) = -\max_{x \in \Omega} (-h)(x) = -\max_{\|x\| \leqslant 1} x(-h) = -\|-h\| =$$
$$= -\|h\| = -h(y) = h(-y),$$

from which the validity of (5.4) easily follows.

3°. In investigating and solving many nonlinear extremal problems, it is also necessary to know the solution of the following auxiliary linear problem:

Let Ω denote a set whose elements are functions that are summable on a bounded subset E of n-dimensional space. Let h denote a function that is measurable and almost everywhere bounded on E. Find an element u of Ω (assumed to exist) such that

$$\int_E \bar{u}(t)\, h(t)\, dt = \min_{u \in \Omega} \int_E u(t)\, h(t)\, dt.$$

We note that the integral

$$h(u) = \int_E u(t)\, h(t)\, dt \tag{5.5}$$

is, for a fixed function h, a linear functional in each of the spaces $L^p(E)$ for $1 \leqslant p \leqslant \infty$. If we consider Ω as a subset of one of these spaces, we may formulate our problem as the problem of finding some element in the set $G_\Omega h$ (or, more generally, as the problem of describing the entire set $G_\Omega h$). Let us show how to solve this problem for certain specific sets Ω.

1) Let Ω denote a sphere of radius c in the space $L^p(E)$, where $(1 < p < \infty)$:

$$\Omega = \left\{ u \in L^p \,\Big|\, \int_E |u|^p\, dt \leqslant c^p \right\}.$$

Since the space L^p is reflexive, we can easily show, by using formulas (5.2) and (5.4), example 5) of subsection 4° of Sec. 3, and the function h in (5.5) as an element of $L^q(E)$ (where $\frac{1}{p} + \frac{1}{q} = 1$) that, in the present case, $G_\Omega h$ consists of a single element $\bar{u}$:

$$\bar{u} = -(\operatorname{sign} h)\left(\frac{|h|}{\|h\|}\right)^{q-1}.$$

2) Let α and β denote members of $L^p(E)$, where $1 < p < \infty$. Consider the set[1]

$$\Omega = \{u \in L^p(E) \mid \alpha \leqslant u \leqslant \beta\}.$$

[1]Here and in what follows, the relations $u \geqslant \alpha$ (resp. $u \leqslant \beta$, resp. $u = \gamma$) should be understood in the sense that, for almost all $t \in E$

$$u(t) \geqslant \alpha(t) \text{ (resp. } (u(t) \leqslant \beta(t)\text{, resp. } u(t) = \gamma(t)).$$

One can easily show that Ω is convex, closed, and bounded in L^p and hence that it is weakly compact. Let us define

$$E_1(h)=\{t\in E \mid h(t)<0\},\quad E_2(h)=\{t\in E \mid h(t)>0\},$$
$$E_3(h)=\{t\in E \mid h(t)=0\}.$$

It follows immediately from (5.5) that the set $G_\Omega h$ consists of functions $\bar{u}$ defined as follows:

$$\bar{u}(t)=\begin{cases}\beta(t) & t\in E_1(h),\\ \alpha(t) & t\in E_2(h),\\ v(t) & t\in E_3(h),\end{cases}$$

where v is an arbitrary element of Ω. The case of greatest interest is the one in which $\alpha(t)=-1$ and $\beta(t)=1$, in other words, the case in which

$$\Omega=\{u \mid |u(t)|\leqslant 1\}.$$

In this case, the general form of elements $\bar{u}$ of $G_\Omega h$ is given by the formula

$$\bar{u}(t)=\begin{cases}-\operatorname{sign} h(t), & \text{if } h(t)\neq 0,\\ v(t)\quad , & \text{if } h(t)=0,\end{cases} \tag{5.6}$$

where v is an arbitrary element of Ω.

3) Suppose that $g\in L^1(E)$ and that $g\geqslant 0$. Let c denote a non-negative constant. Consider the set

$$\Omega=\left\{u\in L^\infty(E) \mid 0\leqslant u\leqslant 1,\ \int_E u(t)\,g(t)\,dt\leqslant c\right\}.$$

The set Ω considered as a subset of the space $L^2([0, T])$ is weakly compact in L^2 and, since h is a linear functional in L^2, it attains a minimum on Ω.

The set $G_\Omega h$ can be described in the present case with the aid of the familiar Neyman-Pearson lemma [3]. We state that lemma without proof. For real k, let us define

$$E_1(k)=\{t\in[0, T] \mid h(t)<kg(t)\},$$
$$E_2(k)=\{t\in[0, T] \mid h(t)>kg(t)\},$$
$$E_3(k)=\{t\in[0, T] \mid h(t)=kg(t)\}.$$

Let k_0 denote the supremum of the set of all nonpositive k satisfying the inequality

$$\int_{E_1(k)} g(t)\,dt \leqslant c.$$

Then, the set $G_\Omega h$ consists of functions $\bar{u}$ defined as follows:

$$\bar{u}(t)=1 \quad (t \in E_1(k_0)),$$
$$\bar{u}(t)=0 \quad (t \in E_2(k_0));$$

$\bar{u}(t)$ is an arbitrary function defined on $E_3(k_0)$ satisfying only the conditions $0 \leqslant \bar{u}(t) \leqslant 1$ and

$$\int_E \bar{u}(t)\,g(t)\,dt = c,$$

$k_0 < 0$ or the conditions $0 \leqslant \bar{u}(t) \leqslant 1$ and

$$\int_E \bar{u}(t)\,g(t)\,dt \leqslant c,$$

if $k_0 = 0$.

Reference [3] gives a generalization of the Neyman-Pearson lemma, which enables us to describe the set $G_\Omega h$ when

$$\Omega = \Big\{ u \in L^\infty(E) \,|\, 0 \leqslant u \leqslant 1, \int_E u(t)\,g_i(t)\,dt = c_i, \\ i=1, \ldots, n;\ g_i \in L^1(E) \Big\}.$$

4) Consider the space $L^p(E)$, where $1 < p < \infty$. Suppose that

$$\Omega = \{ u \in L^p(E) \,|\, u \geqslant 0,\ \|u\| \leqslant 1 \}.$$

For simplicity, in the present case we shall describe not the entire set $G_\Omega h$ but only one element of that set. Let us define

$$E_1 = \{ t \in E \,|\, h(t) \geqslant 0 \},$$
$$E_2 = \{ t \in E \,|\, h(t) < 0 \},$$
$$\Omega_1 = \{ u \in \Omega \,|\, u(t) = 0,\ t \in E_2 \},$$
$$\Omega_2 = \{ u \in \Omega \,|\, u(t) = 0,\ t \in E_1 \}.$$

Then, let us define

$$h^-(t)=\begin{cases}0\ , & t\in E_1\\ h(t), & t\in E_2\end{cases}=\min(h(t),\ 0).$$

Reasoning as in example 1), we can easily show that the function $\bar{u}(t)$ defined by

$$\bar{u}(t)=-\operatorname{sign} h^-(t)\left(\frac{|h^-(t)|}{\|h^-\|}\right)^{q-1}=\begin{cases}0, & t\in E_1\\ \left(\frac{|h^-(t)|}{\|h^-\|}\right)^{q-1}, & t\in E_2,\end{cases}$$

is an element of $G_{\Omega_2}h^-\supset G_{\Omega_2}h$. It immediately follows from the definition of Ω_1 that $0\in G_{\Omega_1}$. By virtue of this, we can easily show that the function $\bar{u}(t)$ defined above is an element of $G_{\Omega}h$.

One can give yet other examples of sets Ω for which the problem of minimizing a linear integral functional can be solved in a comparatively simple manner. In particular, in Sec. 3 of Chapter 4, we shall consider in detail the solution of the problem of minimizing the functional (5.5) on the set

$$\Omega=\left\{u\in L^2([0,\ T])\,\middle|\,|u(t)|\leqslant 1,\ t\in[0,\ T],\ \int_0^T u^2(t)\,dt\leqslant c\quad(0<c<\infty)\right\}.$$

What we have considered in the present section is the problem minimizing the linear integral functional

$$h(u)=\int_0^T\sum_{i=1}^r h_i(t)\,u_i(t)\,dt.$$

on certain convex sets Ω in the space of square-summable, r-dimensional, vector-valued functions $L_r^2([0,\ T])$.

CHAPTER 2

Necessary Conditions for an Extremum

1. The cones $K_x(\Omega)$ and $M_x(\Omega)$. Necessary conditions for an extremum in terms of cones.

1°. In the investigation and study of extremal problems, considerable importance is attached to necessary conditions for an extremum — the "indications of optimality." A knowledge of the necessary conditions enables us to ascertain whether a given element is an extremum of a functional or at least whether it is "suspected" of being an extremum. If no extremum is attained, we can sometimes use the necessary condition to indicate a procedure by which we might "correct" the given element in such a way that it "better" satisfies that condition, in other words, to offer a method of successive approximations for solving a problem.

In recent years, various authors have proposed several methods for obtaining necessary conditions. We mention, for example, [10, 25, 43, 45, 48, 56]. The method that we shall present below is probably not the most general one, but it does appear to us to be simpler than many of the others. This procedure can be used to investigate a broad class of problems that are important in practice, particularly, optimal-control problems (see Chapter 4). Furthermore (and this is very important), by using the necessary conditions obtained by means of this procedure, one can easily construct successive approximations that lead in a certain sense to elements satisfying the necessary conditions (see Chapters 3, 4).

In the general case, the necessary conditions of which we are speaking may be formulated in terms of cones of a special form (denoted respectively by $K_x(\Omega)$ and $M_x(\Omega)$ and providing a "linear approximation" of the set Ω in a neighborhood of a point x). In [25], such cones are called respectively *cones of admissible directions* with respect to restrictions of the inequality type and with respect to restrictions of the equality type.

In what follows, we shall speak almost exclusively of a local extremum and, with rare exceptions, shall not mention this fact explicitly. For definiteness, we shall almost always speak of the necessary conditions for a minimum. The reader will see in the following pages that these conditions can be modified in a natural way for a maximum.

2°. Let X denote a normed space and let Ω denote a subset of X. Let $\overline{\Omega}$ denote, as usual, the closure of Ω. We shall say that an element u is an admissible direction at a point $x \in \overline{\Omega}$ with respect to the set Ω if there exists a number $\alpha_0 > 0$, depending on x and u, such that $x + \alpha u \in \Omega$ whenever $\alpha \in (0, \alpha_0)$.

Obviously, the set of admissible directions is a cone. Let us denote this cone by $K_x(\Omega)$. It follows from the definition of $K_x(\Omega)$ that $0 \in K_x(\Omega)$ if and only if $x \in \Omega$. We note also that $K_x(\Omega)$ is contained in the cone $C(\Omega - x)$ [the conical hull of the set $\Omega - x$]. To see this, suppose that $u \in K_x(\Omega)$. Then, there exists an $\alpha > 0$ such that $x + \alpha u \in \Omega$, from which it follows that

$$u \in \frac{1}{\alpha}(\Omega - x) \subset C(\Omega - x).$$

If Ω is a convex set and $x \in \Omega$, then $K_x(\Omega) = C(\Omega - x)$. To see this, let z denote a member of Ω. Then, for $\alpha \in [0, 1]$, we have $x + \alpha(z - x) \in \Omega$; that is, $z - x \in K_x(\Omega)$.

Thus, $\Omega - x \subset K_x(\Omega)$ and, since $C(\Omega - x)$ is the smallest cone containing $\Omega - x$, it follows that $C(\Omega - x) \subset K_x(\Omega)$ as we wished to show.

As will be clear from what follows, what we shall need is not so much the cone $K_x(\Omega)$ itself as its closure $\overline{K_x(\Omega)}$. Keeping this in mind, let us look at some examples.

1) Let x denote a member of X and let φ denote a functional that is differentiable with respect to directions at the point x. Suppose that $\Omega = \{z \in X \mid \varphi(z) \leqslant \varphi(x)\}$. In this case,

$$\{u \in X \mid \varphi'_x(u) \leqslant 0\} \supset K_k(\Omega) \supset \{u \in X \mid \varphi'_x(u) < 0\}. \tag{1.1}$$

To see this, let u denote a member of X and suppose that $\varphi'_x(u) < 0$. By using the formula on finite increments, we obtain

$$\frac{1}{\alpha}[\varphi(x+\alpha u)-\varphi(x)] = \varphi'_x(u) + \frac{1}{\alpha} o_{x,u}(\alpha), \tag{1.2}$$

where

$$\frac{1}{\alpha} o_{x,u}(\alpha) \xrightarrow[\alpha \to +0]{} 0.$$

There exists an $\alpha_0 > 0$ such that, for $0 < \alpha \leqslant \alpha_0$,

$$\left| \frac{1}{\alpha} o_{x,u}(\alpha) \right| < |\varphi'_x(u)|.$$

Then, it follows from (1.2) that, for these α

$$\varphi(x+\alpha u) < \varphi(x),$$

that is, $x+\alpha u \in \Omega$ and, consequently, $u \in K_x(\Omega)$. Suppose now that $u \in K_x(\Omega)$. Then, for positive α not exceeding some α_0, the difference $\varphi(x+\alpha u) - \varphi(x) \leqslant 0$. Taking the limit in (1.2), we obtain $\varphi_x(u) \leqslant 0$, as we wished to show.

We shall give two specific examples showing that the cone $\bar{K}_x(\Omega)$ can be either open or closed.

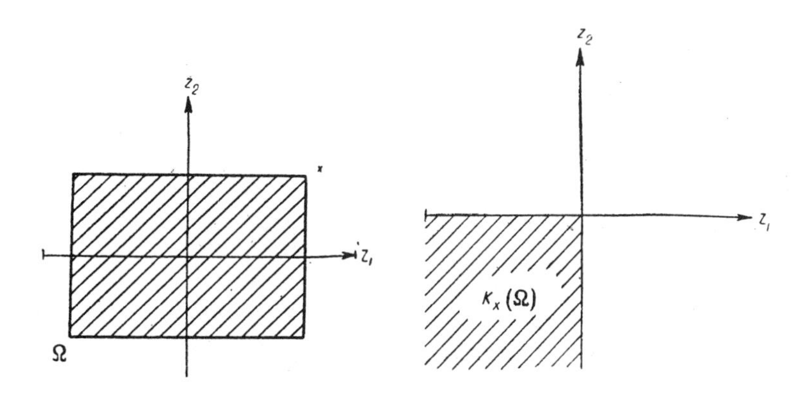

Fig. 1. Fig. 2.

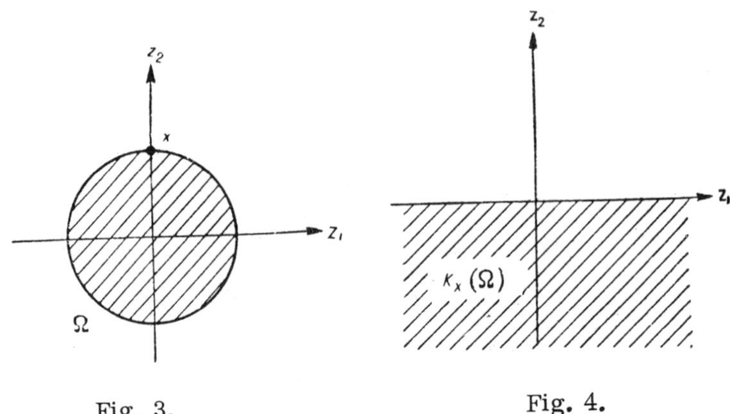

Fig. 3. Fig. 4.

a) $X=R^2$. For $z=(z_1,z_2)\in X$, we have $\varphi(z)=\max(|z_1|,|z_2|)$ and $x=(1,1)$ [see Fig. 1]. One can easily see (see Fig. 2) that the cone $K_x(\Omega)=\{u=(u_1, u_2)\,|\,u_1\leqslant 0,\ u_2\leqslant 0\}$ is closed.

b) As before, suppose $X=R^2$. If $\varphi(z)=z_1^2+z_2^2$, then $x=(1,1)$ (see Fig. 3). In this case, $K_x(\Omega)=\{u=(u_1, u_2)\,|\,u_2<0\}$ is an open cone (see Fig. 4). It is also possible to give an example showing that the cone $K_x(\Omega)$ can fail to be either open or closed.

One can easily show that, if φ is differentiable at a point x (that is, if φ'_x is a linear functional), we have

$$\overline{K_x(\Omega)}=\{u\in X\,|\,\varphi'_x(u)\leqslant 0\}. \tag{1.3}$$

Let us describe a more general class of functionals for which (1.3) holds. We shall say that a functional φ'_x possesses A_1 (resp. property A_2) if, for every $\varepsilon>0$ and every u such that $\varphi'_x(u)=0$, there exists an element $v_1^\varepsilon\in S_\varepsilon(u)$ [resp. an element $v_2^\varepsilon\in S_\varepsilon(u)$] such that $\varphi'_x(v_1^\varepsilon)<0$ (resp. such that $\varphi'_x(v_2^\varepsilon)>0$). Here, as usual,

$$S_\varepsilon(u)=\{v\in X\,|\,\|v-u\|\leqslant\varepsilon\}.$$

If a functional φ'_x is continuous and possesses property A_1, then the closure of the set $\{u\in X\,|\,\varphi'_x(u)<0\}$ coincides with the set $\{u\in X\,|\,\varphi'_x(u)\leqslant 0\}$. Therefore, it follows from (1.1) that equation (1.3) holds.

By reasoning in the same way, we can easily show that if $\Omega'=\{z\in X\,|\,\varphi(z)\geqslant\varphi(x)\}$ and the functional φ_x is continuous and possesses property A_2, then

$$\overline{K_x(\Omega')} = \{u \in X \mid \varphi'_x(u) \geqslant 0\}.$$

An important example of functionals whose derivative with respect to directions at a given point x possesses properties A_1 and A_2 is a functional that is differentiable in the sense of Gateau at a point x and whose gradient at that point is nonzero. To see this, remember that, if a nonzero linear functional vanishes at the center of a sphere, it assumes values of different signs in that sphere.

Let us describe a more general class of functionals whose derivative with respect to directions at a point x possesses properties A_1 and A_2. This class includes functions φ for which φ'_x is a sublinear (resp. superlinear) functional such that

$$\inf_{v \in X} \varphi'_x(v) = -\infty \quad \left[\text{resp. } \left(\sup_{v \in X} \varphi'_x(v) = \infty\right)\right].$$

We shall consider only the case in which φ'_x is a sublinear functional. Suppose that $\varphi'_x(u) = 0$. It follows from our assumption that there exists a $v \in X$ such that $\varphi'_x(v) < 0$. We have, for arbitrary $\varepsilon > 0$,

$$\varphi'_x(u + \varepsilon v) \leqslant \varphi'_x(u) + \varepsilon\varphi'_x(v) = \varepsilon\varphi'_x(v) < 0;$$

that is, φ'_x possesses property A_1. Since φ'_x is a sublinear functional, it follows [see (I.2.7)] that, for arbitrary $z \in X$,

$$\varphi'_x(z) = \max_{h \in \mathfrak{u}_{\varphi'_x}} h(z),$$

where $\mathfrak{u}_{\varphi'_x}$ is the set of all functionals that are supporting for φ'_x. Let f_0 denote a member of $\mathfrak{u}_{\varphi'_x}$ such that

$$0 = \varphi'_x(u) = \max_{u \in \mathfrak{u}_{\varphi'_x}} f(u) = f_0(u).$$

Let us choose $x \in X$ so that $f_0(w) > 0$. Then,[1]

$$\varphi'_x(u+\varepsilon w) = \max_{f \in u'_{\varphi_x}} f(u+\varepsilon w) \geqslant f_0(u+\varepsilon w) = \varepsilon f_0(w) > 0.$$

This proves that φ'_x possesses property A_2.

Thus, if φ is differentiable with respect to directions at a point x and if φ'_x is a sublinear (resp. superlinear) functional such that

$$\inf \varphi'_x(v) = -\infty \ \left(\text{resp. } \varphi'_x(v) = \infty\right),$$
$$\Omega = \{z \in X \mid \varphi(z) \leqslant \varphi(x)\},$$

then,

$$\overline{K_x(\Omega)} = \{u \in X \mid \varphi'_x(u) \leqslant 0\}.$$

2) Suppose that $X = C([a, b])$, where $(-\infty < a < b < \infty)$. Let x denote a member of $C([a, b]$. Define

$$\Omega = \left\{z \in C([a, b]) \mid \max_{t \in [a, b]} z(t) = \max_{t \in [a, b]} x(t)\right\}.$$

$$E(x) = \left\{t \in [a, b] \mid x(t) = \max_{\tau \in [a, b]} x(\tau)\right\},$$
$$E_\delta(x) = \{t \in [a, b] \mid \rho(t, E(x)) < \delta\}.$$

Here,

$$\rho(t, E(x)) = \inf_{\tau \in E(x)} |t - \tau|.$$

Let $u(t)$ denote a function that is continuous on the interval $[a, b]$ and that has the following properties:

$$\text{a) } \max_{t \in E(x)} u(t) = 0; \tag{1.4}$$

b) There exists a $\delta > 0$ such that, for $t \in E_\delta(x)$,

$$u(t) \leqslant 0. \tag{1.5}$$

[1]Actually, we have also proven that, if φ_x is an arbitrary sublinear functional, it possesses property A_2.

Let us show that, in this case, $u \in K_x(\Omega)$. Let us define

$$E^+(u) = \{t \in [a, b] \mid u(t) > 0\}.$$

We may assume without loss of generality that $E^+(u) \neq \varnothing$ because, in the opposite case, the proof is obvious. Define

$$\alpha_0 = \min_{t \in E^+(u)} \left\{ \frac{1}{u(t)} \left[\max_{\tau \in [a, b]} (x(\tau) - x(t))\right] \right\}.$$

Since

$$\max_{t \in E^+(u)} u(t) = \max_{t \in [a, b]} u(t) < \infty$$

and

$$\min_{t \in E^+(u)} \left[\max_{\tau \in [a, b]} x(\tau) - x(t)\right] \geqslant$$

$$\geqslant \min_{t \in [a, b] \setminus E_\delta(x)} \left[\max_{\tau \in [a, b]} x(\tau) - x(t)\right] > 0,$$

we have

$$\alpha_0 \geqslant \min_{t \in E^+(u)} \left[\max(x(\tau) - x(t)) \Big| \max_{t \in E^+(u)} u(t)\right] > 0.$$

It follows from the definition of α_0 that, for $\alpha \in (0, \alpha_0]$ and $t \in E^+(u)$,

$$x(t) + \alpha u(t) \leqslant \max_{\tau \in E} x(\tau).$$

For $t \in [a, b] \setminus E^+(u)$ we have $u(t) \leqslant 0$. Therefore, for arbitrary real α,

$$x(t) + \alpha u(t) \leqslant \max_{\tau \in [a, b]} x(\tau).$$

In particular,

$$\max_{t \in E(x)} [x(t) + \alpha u(t)] = \max_{t \in [a, b]} x(t).$$

Thus, for $\alpha \in (0, \alpha_0]$,

$$\max_{t\in[a,\,b]} (x+\alpha u)(t) = \max_{t\in E(x)} (x+\alpha u)(t) = \max_{t\in E} x(t) = \max_{t\in[a,\,b]} x(t),$$

from which it follows that $u \in K_x(\Omega)$.

In subsection 4°, we shall show that the cone $\overline{K_x(\Omega)}$ coincides with the closure of the set of all functions satisfying conditions (1.4) and (1.5).

3) As before, $X = C([a, b])$; $x \in C([a, b])$, $x \neq 0$;

$$\Omega = \{z = C([a, b]) \mid \|z\| = \|x\|\}.$$

Let us define

$$\begin{aligned}
E^1(x) &= \{t \in [a, b] \mid x(t) = \|x\|\},\\
E^2(x) &= \{t \in [a, b] \mid x(t) = -\|x\|\},\\
E^i_\delta(x) &= \{t \in [a, b] \mid \rho(t, E_i(x)) < \delta\} \quad (i = 1, 2),\\
E^3_\delta(x) &= [a, b] \setminus (E^1_\delta(x) \cup E^2_\delta(x)).
\end{aligned}$$

Let $u(t)$ denote a function that is continuous on $[a, b]$ and satisfies the following conditions:

1) there exists a $\delta > 0$ such that

$$\begin{aligned} u(t) &\leqslant 0, \quad \text{for } t \in E^1_\delta(x)\\ u(t) &\geqslant 0; \quad \text{for } t \in E^2_\delta(x) \end{aligned} \tag{1.6}$$

2) either $\max_{t\in E^1(x)} u(t) = 0$ or $\min_{t\in E^2(x)} u(t) = 0.$ (1.7)

Let us show that in this case, $u \in K_x(\Omega)$. Without loss of generality, we may assume that neither $E^1(x)$ nor $E^2(x)$ is empty (in the opposite case, the proof is merely simplified). We may also assume that δ is sufficiently small that $x(t) > 0$ for $t \in E^1_\delta(x)$ and $x(t) < 0$ for $t \in E^2_\delta(x)$. Define

$$\alpha_0 = \min\left(\min_{t\in E^1_\delta(x)} \left|\frac{2x(t)}{u(t)}\right|,\ \min_{t\in E^2_\delta(x)} \left|\frac{2x(t)}{u(t)}\right|,\ \frac{\|x\| - \|x\|'}{\|u\|}\right),$$

where

$$\|x\|' = \max_{t\in E^3_\delta(x)} |x(t)|.$$

If follows from our assumptions that $\alpha_0 > 0$. For $0 < \alpha \leqslant \alpha_0$, we have the following:

a) If $t \in E_\delta^1(x)$ then $|2x(t)| \geqslant \alpha |u(t)|$. Since $u(t) \leqslant 0$ and $x(t) > 0$, it then follows that

$$0 \geqslant \alpha u(t) \geqslant -2x(t), \quad x(t) \geqslant x(t) + \alpha u(t) \geqslant -x(t),$$
$$|x(t) + \alpha u(t)| \leqslant |x(t)|,$$
$$\max_{t \in E_\delta^1(x)} |x(t) + \alpha u(t)| \leqslant \max_{t \in E_\delta^1(x)} |x(t)| = \|x\|;$$

b) If $t \in E_\delta^2(x)$, in the same way, we obtain

$$\max_{t \in E_\delta^2(x)} |x(t) + \alpha u(t)| \leqslant \|x\|;$$

c) Finally, on the set $E_\delta^3(x)$, we have

$$\max_{t \in E_\delta^3(x)} |x(t) + \alpha u(t)| \leqslant \max_{t \in E_\delta^3(x)} |x(t)| +$$
$$+ \alpha \max_{t \in E_\delta^3(x)} |u(t)| \leqslant \|x\|' + \alpha \|u\| \leqslant \|x\|.$$

Thus,

$$\|x + \alpha u\| = \max_{t \in [a, b]} |x(t) + \alpha u(t)| =$$
$$= \max \Big(\max_{t \in E_\delta^1(x)} |x(t) + \alpha u(t)|, \max_{t \in E_\delta^2(x)} |x(t) + \alpha u(t)|,$$
$$\max_{t \in E_\delta^3(x)} |x(t) + \alpha u(t)| \Big) \leqslant \|x\|.$$

It now follows from (1.7) that $\|x + \alpha u\| = \|x\|$, and this means that $u \in K_x(\Omega)$.

In subsection 4°, we shall show that the closure of the cone $K_x(\Omega)$ coincides with the closure of the cone of all functions satisfying conditions (1.6) and (1.7).

3°. Suppose that f is a functional defined on a normed space X. Let x denote a member of X. Define

$$\Omega_x^f = \{z \in X \mid f(z) < f(x)\}.$$

If the functional f does not attain its absolute minimum at the point x, then $\Omega_x^f \neq \varnothing$. We note that the continuity of f implies that Ω_x^f

contains an interior point. In what follows, we shall denote the cone $K_x(\Omega_x^f)$ by K_x^f. Obviously, $0 \notin K_x^f$.

THEOREM 1.1. *Suppose that the functional f attains a minimum on the set Ω at a point y. Then*

$$K_y(\Omega) \cap K_y^f = \varnothing.$$

Proof: Let us suppose that the theorem is not true and that there exists a $u \in K_y(\Omega) \cap K_y^f$. Since $0 \notin K_y^f$, we have $u \neq 0$. Obviously, the ray

$$\{\alpha u \mid \alpha > 0\} \subset K_y(\Omega) \cap K_y^f.$$

Let α_0 and α_0^f denote numbers such that $y + \alpha u \in \Omega$ for $\alpha \in (0, \alpha_0)$ and $y + \alpha u \in \Omega_y^f$ for $\alpha \in (0, \alpha_0^f)$. Since the minimum of the functional f on Ω is attained at the point y, it follows that $f(y + \alpha u) \geqslant f(y)$ for sufficiently small $\alpha \in (0, \alpha_0^f)$. On the other hand, $f(y + \alpha u) < f(y)$ for $\alpha \in (0, \alpha_0]$ This contradiction proves the theorem.

THEOREM 1.1°. *Suppose that a function f attains its minimum on Ω at a point y and that the cone K_y^f is corporeal. Then,*

$$\overline{K_y(\Omega)} \cap \mathring{K}_y^f = \varnothing.$$

(Here, $\mathring{K}_y^f$ denotes, as usual, the interior of the cone K_y^f.)

Proof: By virtue of Theorem 1.1,

$$K_y(\Omega) \cap K_y^f = \varnothing,$$

and *a fortiori*

$$K_y(\Omega) \cap \mathring{K}_y^f = \varnothing,$$

but then we have

$$\overline{K_y(\Omega)} \cap \mathring{K}_y^f = \varnothing$$

which completes the proof of the theorem.

Theorems 1.1 and 1.1° state necessary conditions for a minimum in terms of the cones $K_y^f(\Omega)$. As Theorem 1.1° shows, if the cone K_y^f is corporeal, it is sufficient to know $\overline{K_y(\Omega)}$.

If we impose certain restrictions on the functional f and the set Ω, we can write the necessary conditions given above in a more convenient form (see Sec. 2 of the present chapter).

4°. The cone $K_x(\Omega)$ of directions that are admissible at the point x with respect to the set Ω approximates linearly the set $\Omega - x$ close to the point 0 (or, what amounts to the same thing, the cone $x + K_x(\Omega)$ with vertex at the point x approximates linearly the set Ω close to the point x). The approximation by means of the cone $K_x(\Omega)$ proves to be sufficiently good for a broad class of sets, for example, for convex sets, for sets that have a nonempty interior (in the case in which x belongs to the closure of the interior), for sets of the form $\varphi(x) = c$, where φ is some functional of the "maximum type," etc. However, for many sets Ω that are important in practice, the approximation by means of the cone $K_x(\Omega)$ proves insufficient. For example, this is the case when $\Omega = \{z \in X \mid \varphi(z) = \varphi(x)\}$, where φ is a strictly convex functional and $K_x(\Omega) = \{0\}$. In connection with this, it becomes necessary to find something that provides a more accurate approximation.

Suppose that $\Omega \subset X$. We shall call on element $u \in X$ an admissible direction in the broad sense of the word at a point $x \in \bar{\Omega}$ with respect to the set Ω if, for every $\varepsilon > 0$, there exist an element $u_\varepsilon \in S_\varepsilon(u)$ and a number $\alpha_\varepsilon \in (0, \varepsilon)$ such that $x + \alpha_\varepsilon u_\varepsilon \in \Omega$.

We denote by $M_x(\Omega)$ the cone of directions that are admissible in the broad sense. Let us look at a few properties of $M_x(\Omega)$.

1) $M_x(\Omega)$ is a closed cone. To see this, suppose that $u_n \to u$, where each $u_n \in M_x(\Omega)$. Let N denote a number such that, for $n > N$, we have $\|u_n - u\| < \frac{\varepsilon}{2}$, where x is an arbitrary positive number. Then, for given ε, there exists an element $u_n^\varepsilon \in S_{\varepsilon/2}(u_n)$ and a number $\alpha_\varepsilon \in \left(0, \frac{\varepsilon}{2}\right]$ such that $x + \alpha_\varepsilon u_n^\varepsilon \in \Omega$. Since

$$\|u_n^\varepsilon - u\| \leqslant \|u_n^\varepsilon - u_n\| + \|u_n - u\| < \varepsilon,$$

the element u is admissible in the broad sense of the word.

2) $0 \in M_x(\Omega)$.

3) $\overline{K_x(\Omega)} \subset M_x(\Omega)$. To see this, note that, if $u \in K_x(\Omega)$, then, for arbitrary $\varepsilon > 0$ we can set $u_\varepsilon = u$ and $\alpha_\varepsilon = \min(\alpha_0, \varepsilon)$, where α_0 is the number appearing in the definition of an admissible direction. Thus, $u \in M_x(\Omega)$ and, consequently, $K_x(\Omega) \subset M_x(\Omega)$. Since $M_x(\Omega)$ is closed, we also have

$$\overline{K_x(\Omega)} \subset M_x(\Omega).$$

Suppose that Ω is a convex set and that $x \in \Omega$. In this case, as one can easily show,

$$M_x(\Omega) = \overline{K_x(\Omega)} = \overline{C(\Omega - x)}.$$

We mention also that, if $\Omega = \{z \in X \mid \varphi(z) \leqslant \varphi(x)\}$, where φ is a functional that is differentiable with respect to directions at the point x, and, if the functional φ'_x is continuous and possesses property A_1, then

$$M_x(\Omega) = \overline{K_x(\Omega)} = \{z \in X \mid \varphi'_x(z) \leqslant 0\}.$$

We shall be interested in the case in which

$$\Omega = \{z \in X \mid \varphi(z) = \varphi(x)\},$$

where φ is a functional that is uniformly differentiable with respect to directions at the point x and whose derivative φ'_x is a continuous functional. Let us show that, in this case,

$$M_x(\Omega) \subset \{u \in X \mid \varphi'_x(u) = 0\}. \quad (1.8)$$

Let u denote a member of $M_x(\Omega)$. Then, for arbitrary $\varepsilon > 0$, there exist an element $u_\varepsilon \in S_\varepsilon(u)$ and a number $\alpha_\varepsilon \in (0, \varepsilon]$ such that

$$\varphi(x + \alpha_\varepsilon u_\varepsilon) = \varphi(x).$$

On the other hand, from the formula for finite increments,

$$\varphi(x + \alpha_\varepsilon u_\varepsilon) = \varphi(x) + \alpha_\varepsilon \varphi'_x(u_\varepsilon) + o_{x, u_\varepsilon}(\alpha_\varepsilon),$$

so that

$$\varphi_x(u_\varepsilon) = -\frac{1}{\alpha_\varepsilon} o_{x, u_\varepsilon}(\alpha_\varepsilon).$$

Since φ is uniformly differentiable, it follows that, for arbitrary $\delta > 0$, there exists $\varepsilon_0 > 0$ such that, for $\varepsilon < \varepsilon_0$,

$$\left| \frac{1}{\varepsilon} o_{x, u_\varepsilon}(\alpha_\varepsilon) \right| < \delta,$$

from which it follows that

$$\varphi'_x(u) = \lim_{\varepsilon \to 0} \varphi'_x(u_\varepsilon) = 0.$$

Let us suppose now that φ is a continuous functional and that φ_x' possesses properties A_1 and A_2. Then,

$$M_x(\Omega)=\{u \in X \mid \varphi_x'(u)=0\}. \tag{1.9}$$

To see that (1.9) is true, let us suppose that it is not, that is, that there exists a u such that $\varphi_x'(u)=0$ but $u \notin M_x(\Omega)$. It follows from this last that there exists an $\varepsilon>0$ such that

$$\varphi(x+\alpha v)-\varphi(x)\neq 0$$

for all $\alpha \in (0, \varepsilon]$ and $v \in S_\varepsilon(u)$. By using the continuity of φ, we can easily show, that for all such α and v, the difference indicated above maintains its sign. To see this, suppose that

$$\varphi(x+\alpha_1 v_1)>\varphi(x); \quad \varphi(x+\alpha_2 v_2)<\varphi(x).$$

Then, by Cauchy's theorem, the function

$$\psi(t)=\varphi((x+\alpha_1 v_1)+t(\alpha_2 v_2-\alpha_1 v_1))-\varphi(x)$$

vanishes at some point $t_0 \in (0, 1)$; that is,

$$\varphi(x+\alpha_1 v_1+t_0(\alpha_2 v_2-\alpha_1 v_1))=\varphi(x),$$

but this is impossible since the element $\alpha_1 v_1+t_0(\alpha_2 v_2-\alpha_1 v_1)$ can be represented in the form αv, where

$$\alpha=t_0\alpha_2+(1-t_0)\alpha_1 \in (0, \varepsilon]$$

and

$$v=\frac{t_0\alpha_2}{t_0\alpha_2+(1-t_0)\alpha_1} v_2+\frac{(1-t_0)\alpha_1}{t_0\alpha_2+(1-t_0)\alpha_1} v_1 \in S_\varepsilon(u).$$

For definiteness, let us suppose that, for all $\alpha \in (0, \varepsilon]$ and $v \in S_\varepsilon(u)$, we have $\varphi(x+\alpha v)>\varphi(x)$. Then, from the formula

$$\varphi(x+\alpha v)=\varphi(x)+\alpha\varphi_x'(v)+o_{x,v}(\alpha)$$

it follows that $\varphi_x'(v)\geqslant 0$ for all $v \in S_\varepsilon(u)$.

This brings us to a contradiction since our assumption was that the functional φ'_x possesses property A_1. A consequence of this is

THEOREM 1.2. *Let φ denote a functional that is continuous on X and uniformly differentiable with respect to directions at a point $x \in X$. Suppose that φ'_x is a sublinear (resp. superlinear) functional and that*

$$\inf_{v \in X} \varphi'_x(v) = -\infty \; (resp. \left(\sup_{v \in X} \varphi'_x(v) = \infty\right)$$

Define $\Omega = \{z \in X \mid \varphi(z) = \varphi(x)\}$. *Then,*

$$M_x(\Omega) = \{u \in X \mid \varphi'_x(u) = 0\}.$$

Theorem 1.2 describes, in particular, the cones $M_x(\Omega)$ for sets Ω defined in terms of the functionals $\max_{t \in [a,b]} x(t)$ [in the space $C([a, b])$]; $\operatorname{vrai\,max}_{t \in [a,b]} x(t)$ (in the space $L^{\infty}([a, b])$], etc.

Let us pause to look in greater detail at the set $\Omega \subset C([a, b])$:

$$\Omega = \left\{z \in C([a, b]) \mid \max_{t \in [a,b]} z(t) = \max_{t \in [a,b]} x(t)\right\}$$

where x is a fixed element of $C([a, b])$, in order to show that in this case, $\overline{K_x(\Omega)} = M_x(\Omega)$.

Let us denote a member of $M_x(\Omega)$. Then, in accordance with Theorem 1.2,

$$\varphi'_x(u) = 0 \text{ resp. } \varphi(z) = \max_{t \in [a,b]} z(t))$$

that is [see (I.3.17)],

$$\max_{t \in E(x)} u(t) = 0,$$

where

$$E(x) = \left\{t \in [a, b],\ x(t) = \max_{\tau \in [a,b]} x(\tau)\right\}.$$

Let x denote a positive number. It follows from the uniform continuity of u that there exists a $\delta > 0$ such that, for $t', t'' \in [a, b]$, where $|t' - t''| \leqslant \delta$, we have $|u(t') - u(t'')| < \varepsilon$.

Let us define

$$E_1^\delta(x) = \{t \in [a, b] \mid \rho(t, E(x)) \leqslant \delta\},$$
$$E_2^\delta(x) = \{t \in [a, b] \mid \rho(t, E(x)) \geqslant 2\delta\},$$
$$E_3^\delta(x) = (a, b) \setminus (E_1^\delta(x) \cup E_2^\delta(x)).$$

One can easily show that the set $E_3^\delta(x)$ is open, that is, that

$$E_3^\delta(x) = \bigcup (c_k, d_k),$$

where $d_k - c_k \leqslant \delta$.

Let us define a function $u_\varepsilon(t)$ on $[a, b]$ by setting

$$u_\varepsilon(t) = \min(0, u(t)), \quad t \in E_1^\delta(x),$$
$$u_\varepsilon(t) = u(t), \quad t \in E_2^\delta(x).$$

If the point a (resp. the point b) does not belong to $E_1^\delta(x) \cup E_2^\delta(x)$, let us set $u_\varepsilon(a) = u(a)$ [resp. $u_\varepsilon(b) = u(b)$]. Finally, if the interval $(c_k, d_k) \subset E_3^\delta(x)$, we shall assume that the function u_ε is linear on that interval. Thus, the function u_ε that we have constructed is continuous. One can easily show that this function satisfies conditions (1.4) and (1.5). Therefore, as was shown in subsection 2°, u_ε belongs to $K_x(\Omega)$. It follows from the definition of δ that

$$\| u - u_\varepsilon \| = \max_{t \in [a, b]} | u(t) - u_\varepsilon(t) | < \varepsilon.$$

Thus,

$$u = \lim_{\varepsilon \to 0} u_\varepsilon$$

from which it follows that, in the present case,

$$\overline{K_x(\Omega)} = M_x(\Omega).$$

By a slightly more complicated reasoning, we can show without great difficulty that, for the set

$$\Omega = \{z \in C([a, b]) \mid \|z\| = \|x\|\} \quad (x \in C([a, b]))$$

we still have

$$\overline{K_x(\Omega)} = M_x(\Omega).$$

5°. If in Theorem 1.1° we replace the cone $\overline{K_{y^*}(\Omega)}$ with the broader cone $M_y(\Omega)$, this theorem ceases to be valid. Let us look at an example bringing this point out.

Suppose that $X=R^2$. We define a functional f on R^2 as follows:

1) If the vector $u=(u_1, u_2)$ lies in the lower closed half-plane ($u_2 \leqslant 0$), we take $f(u)=0$.

2) Suppose that $u_0=(0, 1)$. For elements of the form $\{\alpha u_0\}_{\alpha\geqslant 0}$, we set $f(\alpha u_0)=-\alpha^3$.

3) Suppose that $u=(u_1, 1)\neq u_0$. On the ray $\{\alpha u\}_{\alpha\geqslant 0}$, we define f as follows:

$$f(\alpha u)=\begin{cases}\alpha^2(\alpha-|u_1|), & \alpha\in[0, |u_1|],\\ (\alpha-|u_1|)(4|u_1|-\alpha), & \alpha\in[|u_1|, 4|u_1|),\\ 4|u_1|-\alpha, & \alpha\in[4|u_1|, \infty).\end{cases}$$

We note that the functional f defined in this way is differentiable with respect to directions (though not uniformly so) at the point 0 and that $f_0'=0$. The set

$$\Omega_0^f=\{z\in R^2 \mid f(z)<0=f(0)\}$$

coincides with the open upper half-plane ($u_2>0$). One can easily show that the cone K_0^f constructed at the point 0 with respect to the set Ω_0^f coincides with this half-plane. Let us now define a set Ω as follows (see Fig. 5):

$$\begin{aligned}\Omega=\{v\in R^2 \mid v=\alpha(u_1, u_2);\\ u_1\geqslant 0;\ u_2=1;\\ 2u_1\leqslant\alpha\leqslant 3u_1\}.\end{aligned}$$

Obviously, Ω is a closed corporeal set and the point 0 belongs to the closure of the interior of Ω.

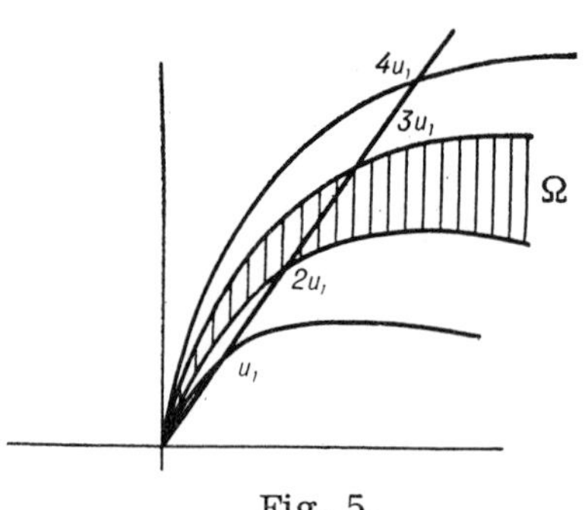

Fig. 5.

It follows from the definition of the functional f that this functional attains its minimum on Ω at the point 0 and only at that point. One can easily see that the cone $K_0(\Omega)$ constructed at the point 0 from the set Ω consists only of the point 0 and that the cone $M_0(\Omega)$ consists of the ray $\{\alpha u_0\}_{\alpha\geqslant 0}$, where $u_0=(0, 1)$. To show, for example, that $u_0\in M_0(\Omega)$, it will be sufficient to note that, for arbitrary $\varepsilon > 0$, the element $u_\varepsilon = \left(\frac{\varepsilon}{2}, 1\right)$ and the number $\alpha_\varepsilon = \varepsilon$ possess the following properties:

$$\|u_0 - u_\varepsilon\| = \frac{\varepsilon}{2} < \varepsilon, \quad \alpha_\varepsilon u_\varepsilon \in \Omega.$$

Thus, in the present case, $M_0(\Omega) \subset \overset{\circ}{K}{}_0^f$.

This shows that replacement of the cone $\overline{K_y(\Omega)}$ with the cone $M_y(\Omega)$ in Theorem 1.1° is inadmissible.

In what follows, we shall frequently assume that the cones K_x^f constructed from the functional f at the point x which we need are corporeal ($\overset{\circ}{K}{}_x^f \neq \varnothing$). To formulate a necessary condition for an extremum in terms of the cone M_y, this is not sufficient, as the example given above shows. Let us show, however, that if we impose certain other requirements on the cone $\overset{\circ}{K}{}_x^f$, the necessary condition ($M_y(\Omega) \cap \overset{\circ}{K}{}_y^f = \varnothing$) that we need will be valid.

Suppose that $\Omega \subset X$, $x \in \overline{\Omega}$ and $\overset{\circ}{K}_x(\Omega) \neq \varnothing$. We shall say that the cone $u' \in \overset{\circ}{K}_x(\Omega)$ is *uniformly open* close to the direction $\overset{\circ}{K}_x(\Omega)$ if there exists a neighborhood $S_\varepsilon(u') \subset \overset{\circ}{K}{}_x^f$, where ε is a number dependent on u', with the following property: for all $u \in S_\varepsilon(u')$, there exists a number α_0 such that, for $\alpha \in (0, \alpha_0]$, we have $x + \alpha u \in \Omega$. If a cone $\overset{\circ}{K}_x(\Omega)$ is uniformly open close to an arbitrary direction $u' \in \overset{\circ}{K}_x(\Omega)$, we shall say that it is uniformly open.

THEOREM 1.3. *Suppose that a functional f attains its minimum on a set Ω at a point x and that the cone $\overset{\circ}{K}{}_y^f = \overset{\circ}{K}_y(\Omega_y^f)$ is nonempty and uniformly open. Then, $M_y(\Omega) \cap \overset{\circ}{K}{}_x^f = \varnothing$.*

Proof: Let us suppose that the theorem is not true and that there exists an element $u' \in M_y(\Omega) \cap \overset{\circ}{K}{}_y^f$. Since $u' \in \overset{\circ}{K}{}_y^f$, there exists a neighborhood $S_\varepsilon(u')$ and a number α_0 with the property that $f(y + \alpha u) < f(y)$ for all $u \in S_\varepsilon(u')$ and $\alpha \in (0, \alpha_0]$. Let η denote a number such that $0 < \eta \leqslant \min(\varepsilon, \alpha_0)$. Since $u' \in M_y(\Omega)$, there exists $u_\eta \in S_\eta(u') \subset S_\varepsilon(u')$ and $\alpha_\eta \in (0, \eta] \subset (0, \alpha_0]$ such that $y + \alpha_\eta u_\eta \in \Omega$. Since the minimum is attained at the point y, we can assume η to be sufficiently small that

$$f(y + \alpha_\eta u_\eta) \geqslant f(y).$$

This contradiction completes the proof of the theorem.

We point out that, if we impose on the functional f and the set Ω various restrictions that are natural in practical problems, we can write the necessary condition for a minimum contained in Theorem 1.3 in a more convenient form (see Sec. 2 of this chapter).

It is easy to show that the cone $\overset{\circ}{K}{}_0^f$ that we considered in our example is nonuniformly open. Let us give an important example of a uniformly open cone.

Let f denote a functional that is differentiable with respect to directions at a point x and suppose that the functional f'_x is continuous. Reasoning as in the examination of example 1) in subsection 1°, we can easily show [see (1.1)] that

$$\{u \in X \mid f'_x(u) \leqslant 0\} \supset K_x^f \supset \{u \in X \mid f'_x(u) < 0\}.$$

Since, by virtue of the continuity of f'_x, the set

$$\{u \in X \mid f'_x(u) < 0\} = (f'_x)^{-1}[(0, -\infty)]$$

is open, we have

$$\{u \in X \mid f'_x(u) \leqslant 0\} \supset \overset{\circ}{K}{}_x^f \supset \{u \in X \mid f'_x(u) < 0\}. \tag{1.10}$$

Let us suppose now that the functional f'_x possesses A_2: if $f'_x(u) = 0$, then, for arbitrary $\varepsilon > 0$, there exists a $v^\varepsilon \in S_\varepsilon(u)$ such that $f'_x(v^\varepsilon) > 0$.

Obviously, $v^\varepsilon \notin K_y^f$, because

$$\frac{1}{\alpha}(f(x + \alpha v^\varepsilon) - f(x)) = f'_x(v^\varepsilon) + \frac{1}{\alpha} o_{x, v^\varepsilon}(\alpha) > 0$$

for sufficiently small α. Thus, in an arbitrary neighborhood of the point $u \in (f'_x)^{-1}(0)$, there is an element that does not belong to K_x^f and hence [see (1.10)]

$$\overset{\circ}{K}{}_x^f = \{u \in X \mid f'_x(u) < 0\}. \tag{1.11}$$

Let us show now that, if f uniformly differentiable with respect to directions at a point x and if f'_x is a continuous functional possessing property A_2, then the cone $\overset{\circ}{K}{}_x^f$ is uniformly open.

Let us take $u' \in \overset{\circ}{K}{}_x^f$. It follows from (1.11) that $f'_x(u') < 0$. Since f'_x is a uniform derivative, there exist an $\varepsilon > 0$ and an α_0 such that, for all $u \in S_\varepsilon(u')$ and $\alpha \in (0, \alpha_0]$,

$$\left|\frac{1}{\alpha} o_{x,u}(\alpha)\right| < \frac{1}{2}\left|f'_x(u')\right|.$$

Since f'_x is a continuous functional, we may assume that the neighborhood $S_\varepsilon(u')$ is chosen in such a way that, for $u \in S_\varepsilon(u')$, we have

$$f'_x(u) < f'_x(u') + \frac{1}{2}|f'_x(u')|.$$

We now have, for $u \in S_\varepsilon(u')$ and $\alpha \in (0, \alpha_0]$,

$$\frac{1}{\alpha}(f(x+\alpha u) - f(x)) = f'_x(\alpha) + \frac{1}{\alpha} o_{x,u}(\alpha) <$$
$$< f'_x(u) + \frac{1}{2}|f'_x(u')| + \frac{1}{2}|f'_x(u')| < 0,$$

from which it follows that $x + \alpha u \in \Omega^f_x$. Thus, all elements u of $S_\varepsilon(u')$ are dependent only on m which proves the uniform openness of $\overset{\circ}{K}{}^f_x$.

As was shown above (cf. footnote on p. 70), the sublinearity of the functional f'_x implies that that functional possesses property A_2. Thus, we have, in particular, proven

THEOREM 1.4. *If f is uniformly differentiable with respect to directions at a point x and if f'_x is a sublinear functional, then the cone $\overset{\circ}{K}{}^f_x$ is uniformly open.*

2. Necessary conditions for a minimum of certain classes of functionals and sets.

1°. In the first two subsections of this section, we shall consider necessary conditions for a minimum for functionals that are differentiable with respect to directions. Here, we shall always assume, without always mentioning it, that the derivative is a continuous functional. The conditions that we shall need can be obtained as a consequence of the general Theorems 1.1 and 1.3. However, we shall give a complete proof of them.

THEOREM 2.1. *Suppose that a functional f attains a minimum on the set Ω at a point y and that it is differentiable with respect to directions at that point. Then,*

$$\min_{u \in K_y(\Omega)} f'_y(u) = 0. \tag{2.1}$$

Proof: Let u denote a member of $K_y(\Omega)$. Then, since the minimum is attained at the point y, we have for sufficiently small α,

$$\frac{1}{\alpha}[f(y+\alpha u) - f(y)] = f'_y(u) + \frac{1}{\alpha} o_{y,u}(\alpha) \geqslant 0,$$

from which it follows that $f'_y(u) \geqslant 0$. On the other hand, remembering that $0 \in K_y(\Omega)$, we have

$$\min_{u \in K_y(\Omega)} f'_y(u) \leqslant f'_y(0) = 0.$$

This completes the proof of the theorem.

Remark. The continuity of f'_y implies that, under the conditions of the theorem,

$$\min_{u \in \overline{K_y(\Omega)}} f'_y(u) = 0. \tag{2.2}$$

2°. Let us now look at the case when f'_y is a uniform derivative.

THEOREM 2.2. *Suppose that a functional f attains a minimum on the set Ω at a point y and is uniformly differentiable with respect to direction at that point. Then,*

$$\min_{u \in M_y(\Omega)} f'_y(u) = 0. \tag{2.3}$$

Proof: Let us suppose that there exists a $u_0 \in M_y(\Omega)$ such that $f'_y(u_0) = -\rho < 0$. By virtue of the continuity of f'_y, there exists a $\eta > 0$ such that $f'_y(u) < -\frac{\rho}{2}$ for $u \in S_\eta(u_0)$. For $\alpha > 0$ and $u \in S_\eta(u_0)$,

$$f(y+\alpha u) = f(y) + \alpha f'_y(u) + o_{y,u}(\alpha) < f(y) + \alpha\left(-\frac{1}{2}\rho\right) + o_{y,u}(\alpha).$$

The uniform differentiability of f implies that there exist $\delta>0$ and $\alpha_0>0$ such that, for $u\in S_\delta(u_0)$ and $\alpha\in(0, \alpha_0]$,

$$\left|\frac{1}{\alpha} o_{y,u}(\alpha)\right|<\frac{1}{4}\rho.$$

Suppose that $0<\varepsilon\leqslant\min(\eta, \delta, \alpha_0)$. Since $u_0\in M_y(\Omega)$, there exist and $\alpha_\varepsilon\in(0, \varepsilon]$ such that $y+\alpha_\varepsilon u_\varepsilon\in\Omega$. We have

$$f(y+\alpha_\varepsilon u_\varepsilon)<f(y)+\alpha\left(-\frac{1}{2}\rho\right)+o_{y,u_\varepsilon}(\alpha_\varepsilon)<$$
$$<f(y)+\alpha\left(-\frac{1}{2}\rho\right)+\frac{1}{4}\alpha\rho=f(y)-\frac{1}{4}\alpha\rho<f(y).$$

On the other hand, since a (relative) minimum of f on Ω is attained at the point y, we see that, for sufficiently small ε,

$$f(y+\alpha_\varepsilon u_\varepsilon)\geqslant f(y).$$

This contradiction proves the theorem.

3°. Let f denote a convex functional. Then, it follows from Theorem 3.1 of Chapter 1 that f is differentiable with respect to directions at every point.

THEOREM 2.3. *Let f denote a convex functional defined on a space X, let K denote a cone in X, and let y denote a point in X at which*

$$\min_{u\in K} f'_y(u)=0.$$

Then,

$$\min_{u\in K} f(y+u)=f(y).$$

Proof: Let us suppose that the assertion of the theorem is untrue. Then, there exists a $u_0\in K$ such that $f(y+u_0)<f(y)$. We have

$$f(y+u_0)=f(y)+f'_y(u_0)+o_{y,u_0}(1).$$

By virtue of Theorem 3.1 of Chapter 1, the function $o_{y,u_0}(\alpha)$ is nonnegative. Therefore,

$$f'_y(u_0)=f(y+u_0)-f(y)-o_{y,u_0}(1)<0,$$

which contradicts the condition of the theorem. This completes the proof of the theorem.

COROLLARY. *Let f denote a convex functional defined on X, let Ω denote a subset of X, and let y denote a point in Ω at which*

$$\min_{u \in \overline{K_y(\Omega)}} f'_y(u) = 0 \quad \left(\textit{resp.} \min_{u \in M_y(\Omega)} f'_y(u) = 0\right)$$

Let $S_\varepsilon(y)$ denote a neighborhood of y such that

$$\begin{gathered}(y + K_y(\Omega)) \cap \Omega \supset S_\varepsilon(y) \cap \Omega, \\ (\text{resp. } y + M_y(\Omega)) \cap \Omega \supset S_\varepsilon(y) \cap \Omega.\end{gathered} \tag{2.4}$$

Then, f attains a local minimum on Ω at the point y.

On the other hand, if we assume instead of the condition (2.4), the stronger condition

$$y + \overline{K_y(\Omega)} \supset \Omega \qquad \text{or} \qquad y + M_y(\Omega) \supset \Omega,$$

then f attains its absulute minimum on Ω at the point y.

4°. Let p denote a sublinear functional. It follows from the remark to Theorem 3.2 of Chapter 1 that p is uniformly differentiable and its derivative p'_y is a sublinear functional.

Thus, in testing for a minimum of a sublinear functional, one can always apply Theorem 2.2 (which, in view of the fact that $M_y(\Omega) \supset \overline{K_y(\Omega)}$, provides a stronger necessary condition).

Let us transform condition (2.3) in the present case. We have

$$\begin{gathered}0 = \min_{u \in M_y(\Omega)} p'_y(u) = \min_{v \in M_y(\Omega)+y} p'_y(v-y) \leqslant \min_{v \in M_y(\Omega)+y} (p'_y(v) + \\ + p'_y(-y)) = \min_{v \in M_y(\Omega)+y} p'_y(v) + p'_y(-y),\end{gathered}$$

from which it follows that

$$-p'_y(-y) \leqslant \min_{v \in M_y(\Omega)+y} p'_y(v).$$

Let us evaluate $-p'_y(-y)$:

$$\begin{gathered}-p'_y(-y) = -\lim_{\alpha \to +0} \frac{1}{\alpha}(p(y - \alpha y) - p(y)) = \\ = -\lim_{\alpha \to +0} \frac{1}{\alpha}((1-\alpha)p(y) - p(y)) = p(y).\end{gathered}$$

Remembering that $y \in M_y(\Omega)+y$ and noting that $p(y)=p'_y(y)$, we obtain

$$\min_{v \in M_y(\Omega)+y} p'_y(v)=p(y),$$

or, what amounts to the same thing,

$$\min_{u \in M_y(\Omega)} p'_y(u+y)=p(y). \tag{2.5}$$

5°. Here, we shall assume that the set Ω, the point y, and the functional f (which attains its minimum on Ω at point y) are such that the cones $K_y(\Omega)$, $M_y(\Omega)$, and $\mathring{K}_y^f$ are also convex and, in addition, that $\mathring{K}_y^f \neq \varnothing$. In what follows, the following theorem will play an important role.

SEPARABILITY THEOREM (Eydel'gayt). *Let A_1 and A_2 denote convex sets in a normed space. Suppose that $\mathring{A}_1 \neq \varnothing$ and $\mathring{A}_1 \cap A_2 = \varnothing$. Then, there exists a linear functional $h \neq 0$ such that*

$$\sup_{x \in \mathring{A}_1} h(x) \leqslant \inf_{x \in A_2} h(x).$$

By applying this theorem and Theorems 1.1 and 1.3 we can easily establish the validity of the following necessary conditions for a minimum:[1]

THEOREM 2.4. *Suppose that a functional f attains its minimum on Ω at a point y. Let $K_y(\Omega)$ and $\mathring{K}_y^f$ denote convex cones. Suppose that $\mathring{K}_y^f$ is nonempty. Then, there exists a linear functional $h \neq 0$ such that*

$$\min_{u \in \overline{K_y(\Omega)}} h(u) = \sup_{u \in K_y^f} h(u) = 0. \tag{2.6}$$

[1]The overall scheme of applying the separability theorems to the investigation of extremal problems was apparently first used in the works of L. V. Kantorovich [31] and M. G. Krein [37].

THEOREM 2.5. *Suppose that a functional f attains its minimum on Ω at a point y. Let $\overset{\circ}{K}{}^f_y$ denote a nonempty uniformly open convex cone, and let $M_y(\Omega)$ denote a convex cone. Then, there exists a nonzero $h \in X^*$ such that*

$$\min_{u \in M_y(\Omega)} h(u) = \sup_{u \in K^f_y} h(u) = 0. \tag{2.7}$$

Let us show, that by imposing certain restrictions on the functional f, we can rewrite the equation

$$\sup_{u \in K^f_y} h(u) = 0, \tag{2.8}$$

which appears in (2.6) and (2.7) in a more convenient form.

6°. Let f denote a quasiconvex functional. Then, the set Ω^f_y is convex. Therefore, as we noted in Sec. 1,

$$K^f_y = K_y(\Omega^f_y) = C(\Omega^f_y - y),$$

where $C(\Omega^f_y - y)$ is the conical hull of $\Omega^f_y - y$. Obviously, in this case,

$$0 = \sup_{u \in K^f_y} h(u) = \sup_{u \in C(\Omega^f_y - y)} h(u) \geqslant \sup_{u \in \Omega^f_y - y} h(u) = \sup_{x \in \Omega^f_y} h(x - y) = 0.$$

Condition (2.8) is, in this case, equivalent to the condition

$$\sup_{x \in \Omega^f_y} h(x) = h(y). \tag{2.9}$$

7°. Let p denote a sublinear functional. Since a sublinear functional is convex and *a fortiori* quasiconvex, (2.9) is valid for it.

It follows from Theorem 2.8 of Chapter 1 that condition (2.9) is equivalent in the present case to the condition that there exists a $\lambda > 0$ such that $\lambda h \in U^y_p$. In what follows, we shall be interested in the functional h which appears in (2.9) only up to a positive constant factor. Therefore, we may say without loss of generality that, in the present case, (2.9) is equivalent to the condition that $h \in u_p^y$.

8°. Let p denote a seminorm. It follows from Remark 3 following Theorem 3.3 of Chapter 1 that

$$u_p^y = \{f \in X^* \mid \|f\|_p = 1,\ f(y) = p(y)\},$$

where

$$\|f\|_p = \sup_{p(x) \leqslant 1} |f(x)|.$$

Thus, we may say without loss of generality that (2.9) is then equivalent to the condition

$$\|h\|_p = 1,\ h(y) = p(y).$$

In particular, if p is a norm, then the functional satisfies (2.9) if and only if

$$\|h\| = 1;\ h(y) = p(y).$$

9°. Let f denote a continuous functional defined on X, let K denote a cone in X, and let y denote a point in X such that

$$\Omega_y^f \neq \varnothing.$$

Suppose that there exists an $h \in X^*$ such that

$$\min_{u \in K} h(y + u) = h(y) = \sup_{x \in \Omega_y^f} h(x).$$

Since the set $\Omega_y^f = f^{-1}[(f(y), -\infty)]$ is open, it follows from the equations given above that

$$\Omega_y^f \cap (y + K) = \varnothing.$$

It now follows from the definition of Ω_y^f that f attains its absolute minimum on the set $y + K$ at the point y.

Suppose now that f is a quasiconvex continuous functional defined on X. Let Ω denote a subset of X. Suppose that there exists a point $y \in \Omega$ such that

$$\min_{u \in \overline{K_y(\Omega)}} h(u+y) = h(y) = \sup_{u \in K_y^f} h(u+y),$$

$$\left(\text{resp.} \quad \min_{u \in M_y(\Omega)} h(u+y) = h(y) = \sup_{u \in K_y^f} h(u+y)\right).$$

Then, in view of (2.9), the relationships given above can be written as follows:

$$\min_{u \in \overline{K_y(\Omega)}} h(u+y) = h(y) = \sup_{x \in \Omega_y^f} h(x)$$

$$\left(\text{resp.} \quad \min_{u \in M_y(\Omega)} (u+y) = h(y) = \sup_{x \in \Omega_y^f} h(x)\right).$$

If there exists a neighborhood $S_\varepsilon(y)$ of the point y such that

$$(y + \overline{K_y(\Omega)}) \cap \Omega \supset S_\varepsilon(y) \cap \Omega$$

$$(\text{resp.} \ ((y + M_y(\Omega)) \cap \Omega \supset S_\varepsilon(y) \cap \Omega),$$

then f attains a local minimum on Ω at the point y. On the other hand, if

$$y + \overline{K_y(\Omega)} \supset \Omega$$

$$(\text{resp.} \quad y + M_y(\Omega) \supset \Omega),$$

then f attains its absolute minimum on Ω at the point y.

10°. Let Ω denote a closed set. Then, as was mentioned in Sec. 1, the following relationships are valid:

$$M_y(\Omega) = \overline{K_y(\Omega)}, \ K_y(\Omega) = C(\Omega - y).$$

It follows from the coincidences of the cones $\overline{K_y(\Omega)}$ and $M_y(\Omega)$ that it is not sensible in the present case to apply Theorems 2.2 and 2.5. The equation $K_y(\Omega) = C(\Omega - y)$ enables us to simplify considerably the necessary conditions (2.2) and (2.6) [in much the same way as we did in subsection 6° for the convex set Ω_y^f]. Specifically, let g denote a functional that is equal to 0 at the point 0. Then, since $C(\Omega - y) \supset \Omega - y$, it follows from the condition

$$\min_{u \in C(\Omega - y)} g(u) = 0$$

that

$$0 = \min_{u \in C(\Omega - y)} g(u) \leqslant \min_{u \in \Omega - y} g(u) = \min_{x \in \Omega} g(x - y) \leqslant 0.$$

Thus, the relation

$$\min_{u \in \overline{K_y(\Omega)}} g(u) \leqslant \min_{u \in K_y(\Omega)} g(u) = 0$$

is equivalent to the relation

$$\min_{x \in \Omega} g(x - y) = 0. \tag{2.10}$$

Keeping the results of subsections 1° - 9° of the present section and formula (2.10) in mind, we can easily formulate necessary conditions for a minimum in the case of a convex set. (For a convex functional, these conditions are also sufficient.)

Let f denote a functional defined on a normed space X. Suppose that it attains its minimum on a convex subset Ω at a point y. Then, we have the following:

1) If f is differentiable with respect to directions at a point y, it is necessary that

$$\min_{x \in \Omega} f'_y(x - y) = 0. \tag{2.11}$$

If, in addition, f is a convex functional, then (2.11) is a necessary and sufficient condition for an absolute minimum of f on Ω.

Remark. If f is a differentiable in the sense of Gateau at a point y, condition (2.11) is equivalent to the condition

$$\min_{x \in \Omega} Fy(x) = Fy(y). \tag{2.12}$$

2) If p is a sublinear functional, then a necessary and sufficient condition for an absolute minimum is that

$$\min_{x \in \Omega} p'_y(x) = p(y). \tag{2.13}$$

3) If $\overset{\circ}{K}{}^f_y$ is a nonempty convex cone, it is necessary that there exist a nonzero $h \in X^*$ such that

$$\min_{x \in \Omega} h(x - y) = \sup_{u \in K^f_y} h(u) = 0. \tag{2.14}$$

4) If f is a quasiconvex functional, a necessary and sufficient condition for it to have an absolute minimum at a point y is that there exist a nonzero $h \in X^*$ such that

$$\min_{x \in \Omega} h(x) = h(y) = \sup_{x \in \Omega^f_y} h(x). \tag{2.15}$$

5° If p is a sublinear functional, a necessary and sufficient condition for an absolute minimum is that there exist an $h \in U^y_p$ such that

$$\min_{x \in \Omega} h(x) = h(y). \tag{2.16}$$

6° If p is a seminorm, a necessary and sufficient condition for an absolute minimum is that there exist an $h \in X^*$ such that

$$\|h\|_p = 1, \ h(y) = p(y) = \min_{x \in \Omega} h(x), \tag{2.17}$$

where

$$\|h\|_p = \sup_{p(x) \leqslant 1} |h(x)|.$$

In particular, if p is a norm, then (2.17) is equivalent to the condition

$$\|h\| = 1, \ h(y) = \|y\| = \min_{x \in \Omega} h(x). \tag{2.18}$$

11°. We now present a necessary condition for a functional f that is differentiable with respect to directions to have a minimum on the entire space X. Remembering that X is a convex set and applying formula (2.11), we see that, if the functional f attains a minimum on X at the point y, then

$$\min_{u \in X} f'_y(u) = 0. \tag{2.19}$$

If f is a convex functional, then (2.19) is a sufficient condition for f to have an absolute minimum on X at the point y.

Let us stop to consider the special case in which f is differentiable in the sense of Gateau at a point y. In this case, (2.19) is equivalent to the condition

$$Fy = 0, \tag{2.20}$$

where F is, as usual, the gradient of the functional f.

12°. Let Ω denote a convex weakly compact subset of a Banach space X. Let f denote a function that attains it minimum on Ω at a point y and is differentiable in the sense of Gateau at that point. Then, the necessary condition (2.12) holds. In the case in which $Fy \neq 0$, this condition means that $y \in G_\Omega Fy$ [where G_Ω is the mapping defined by formula (I.5.1)]. In other words, in this case, (2.12) means that the hyperplane

$$H = \{x \in X \mid Fy(x) = Fy(y)\}$$

is a supporting hyperplane for the set Ω at the point y (see Fig. 6).

Let us suppose that f is differentiable on Ω. Suppose that for every nonzero member h of X^*, the set $G_\Omega h$ consists of a single element (which we also denote by $G_\Omega h$). Under these assumptions, the point y satisfies condition (2.12) if and only if it is a solution of one or the other of the following two equations:

$$Fx = 0, \quad G_\Omega Fx = x. \tag{2.21}$$

Here, if y is a solution of the first of equations (2.21), it follows from (2.20) that the necessary condition for f to have its minimum over the entire space X at the point y is satisfied. We note that, if f is a lower-semicontinuous functional, it follows from Theorem 4.1 of Chapter 1 that it attains its minimum over Ω. Therefore, in this case, one of the two equations (2.21) must necessarily have a solution belonging to Ω. Thus, the necessary condition (2.12) can be used to investigate certain classes of certain nonlinear equations. For this purpose, we may apply the necessary condition (2.3). (The question of investigating nonlinear equations with the use of (2.3) and (2.12) is considered in detail in [52, 53].)

In conclusion, we note that, when Ω is a convex weakly compact set, the necessary conditions (2.14)-(2.18) can also be written in terms of the mapping G_{Ω}.

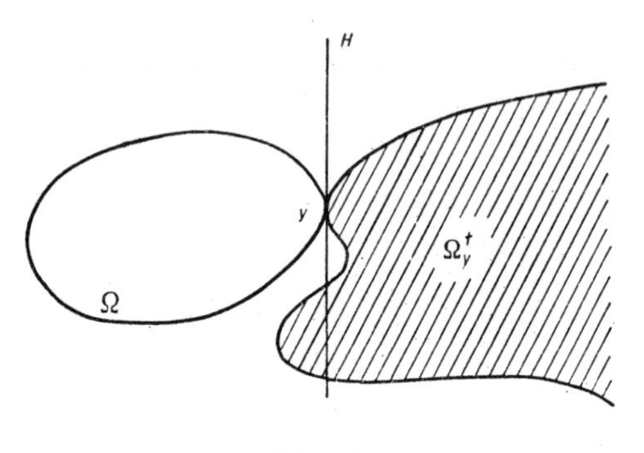

Fig. 6.

13°. Suppose that a functional f attains its minimum on a set Ω at a point y, and that it is differentiable in the sense of Gateau at that point. Then, if the cone $K_y(\Omega)$ is a linear set, the necessary condition (2.1) may be rewritten as follows: for arbitrary $u \in K_y(\Omega)$,

$$Fy(u) = 0. \qquad (2.22)$$

To see this note that, if there were a $u_0 \in K_y(\Omega)$ such that $Fy(u_0) \neq 0$, we would have

$$\inf_{u \in K_y(\Omega)} Fy(u) \leqslant \inf_{-\infty < \lambda < \infty} Fy(\lambda u_0) = -\infty,$$

which contradicts (2.1). The converse assertion [that (2.22) implies (2.1)] is obvious. In particular, if Ω is a linear manifold of the form $z + \Omega'$ (where Ω' is a linear set), then condition (2.22) is equivalent to the following condition: for arbitrary $u \in \Omega'$,

$$Fy(u) = 0.$$

In the present case ($K_y(\Omega)$ linear), the necessary conditions (2.16) - (2.18) can also be represented in a form analogous to (2.23).

Thus, for example, condition (2.16) is equivalent to the following one: there exists an $h \in U_p^y$ such that, for $u \in \Omega'$,

$$h(u) = 0. \tag{2.24}$$

Now, suppose that f is a functional that attains its minimum over the set Ω at a point y and that f is differentiable in the sense of Fréchet at that point. Let us suppose that the cone $M_y(\Omega)$ is a linear set. It then follows from (2.3) that, for $u \in M_y(\Omega)$,

$$Fy(u) = 0. \tag{2.25}$$

Of special interest is the case in which

$$\Omega = \{x \in X \mid \varphi(x) = C\},$$

where φ is a functional that is differentiable in the sense of Fréchet at a point y. Let us denote the gradient of φ at the point y by Φy and let us assume that $\Phi y \neq 0$. Then (see Theorem 1.2), $M_y(\Omega) = \Phi y)^{-1}(0)$. In the case $Fy \neq 0$, it follows from (2.25) that the hyperplanes of the functionals Fy and Φy coincide, and this means that Fy is proportional to Φy. Thus, we have proven the following theorem [6]:

THEOREM (Lyusternik). *Suppose that functionals f and φ are defined on normed space X and that f attains a minimum (at least a local one) on a nonempty set $\Omega = \{x \in X \mid \varphi(x) = c\}$ at a point y. Then, if f and φ are differentiable in the sense of Fréchet at the point y and if their gradients Fy and Φy are nonzero at that point, there exists a $\mu \neq 0$ such that*

$$Fy = \mu \Phi y.$$

Let us give a generalization of Lyusternik's theorem. Suppose as before that $\Omega = \{x \in X \mid \varphi(x) = c\}$ and that f attains a minimum on Ω at a point y. We shall assume that φ is differentiable in the sense of Fréchet at the point y, that $\Phi y \neq 0$, and that the functional f is such that $\overset{\circ}{K}{}_y^f$ is a uniformly open convex cone. Reasoning as in the proof of Lyusternik's theorem and using Theorem 2.5, we can easily show that for some $\mu \neq 0$,

$$\sup_{u \in K_y^f} \mu\Phi y(u) = 0.$$

Considering separately the cases $\mu < 0$ and $\mu > 0$, we see that the point y must satisfy one of the two equations

$$\inf_{u \in K_y^f} \Phi y(u) = 0, \quad \sup_{u \in K_y^f} \Phi y(u) = 0.$$

If f is a sublinear functional, the condition

$$\sup_{u \in K_y^f} \mu\Phi y(u) = 0$$

is equivalent to the condition (see subsection 7°) that

$$\mu\Phi y \in \boldsymbol{u}_f^y.$$

14°. Let us look at the following important example, which we shall need later. Suppose that X is a normed space, that A is continuously differentiable operator mapping X into $C(E)$, where E is a closed bounded subset of a finite-dimensional space, and that a functional f is defined on X by the formula: for $x \in X$,

$$f(x) = \max_{t \in E} (Ax)(t)$$

Since the functional p defined on the space $C(E)$ by

$$p(y) = \max_{t \in E} y(t),$$

is continuously differentiable with respect to directions at an arbitrary point $y \in C(E)$ and its derivative p'_y is a continuous functional. (We recall that

$$u \in C(E) p'_y(u) = \max_{t \in E(y)} u(t),$$

where

$$E(y) = \{t \in E \mid y(t) = \max_{\tau \in E} y(\tau)\})$$

for $u \in C(E)$.) Therefore, the function f is differentiable with respect to directions at an arbitrary point $x \in X$; also [see formula (I.3.31)], for arbitrary $u \in X$,

$$f'_x(u) = \max_{t \in E(Ax)} (A'_x u)(t).$$

It is easy to verify that the functional f'_x is continuous. If Ω is a subset of X, then a necessary condition for f to attain a minimum on Ω at a point y can be written in the form [see (2.2)]

$$\min_{u \in \overline{K_y(\Omega)}} \max_{t \in E(Ay)} (A'_y u)(t) = 0. \tag{2.26}$$

If Ω is a convex set, condition (2.26) is equivalent to the condition [see (2.11)]

$$\min_{x \in \Omega} \max_{t \in E(Ay)} \left(A'_y(x-y)\right)(t) = 0. \tag{2.27}$$

Finally, if Ω coincides with all X, then (2.27) is equivalent [see (2.19)] to the condition

$$\min_{x \in X} \max_{t \in E(Ay)} \left(A'_y(x)\right)(t) = 0. \tag{2.28}$$

15°. All these necessary conditions for a minimum can be carried over with natural modifications to the case of a maximum. Thus, for example, we have the following theorem (analogous to Theorem 2.2):

THEOREM 2.2′. *Suppose that a functional f attains a maximum on a set Ω at a point y and that f is uniformly differentiable with respect to directions at that point. Then*

$$\max_{u \in M_y(\Omega)} f'_y(u) = 0. \tag{2.3′}$$

The analog of condition (2.11) consists in the following: let f denote a functional defined on a normed space X and suppose that f attains a maximum on a convex set Ω at a point y. Then, if f is differentiable with respect to directions at that point, we necessarily have

$$\max_{x \in \Omega} f'_y(x-y) = 0. \tag{2.11'}$$

If, in addition, f is a concave functional, then (2.11′) is a necessary and sufficient condition for f to have an absolute maximum on Ω.

Let us give the analog of condition (2.14). Suppose that a functional f is defined on X. Let x denote a member of X. Let us define

$$\tilde{\Omega}_x^f = \{z \in X \mid f(z) > f(x)\},$$

and

$$\tilde{K}_x^f = K_x\left(\tilde{\Omega}_x^f\right).$$

Suppose now that f attains a maximum on a concave set Ω at a point y. Then, if $\tilde{K}_y^f$ is a convex cone with nonempty interior, there exists a nonzero $h \in X^*$ such that

$$\max_{x \in E} h(x-y) = \inf_{u \in \tilde{K}_y^f} h(u) = 0. \tag{2.14'}$$

16°. In conclusion, let us give a geometrical interpretation of the necessary condition for a maximum in one particular case. Suppose that functional f is defined on n-dimensional space R^n by

$$f(x) = \min_{t \in E} g(x, t),$$

where $x \in R^n$, where E is a closed bounded subset of m-dimensional R^m and where g is a function defined on the set $R^n \times E$ that is continuous on that set and has, for arbitrary $x \in R^n$, a continuous partial derivative with respect to x

$$\frac{\partial g}{\partial x} = \left(\frac{\partial g}{\partial x_1}, \ldots, \frac{\partial g}{\partial x_n}\right).$$

The functional f is differentiable with respect to directions at an arbitrary point $x \in R^n$: for arbitrary $u \in R^n$, we have [compare with (I.3.34)]

$$f'_x(u) = \min_{t \in E(x)} \left(\frac{\partial g}{\partial x}(x, t) \right)(u),$$

where

$$E(x) = \left\{ t \in E \mid g(x, t) = \min_{\tau \in E} g(x, \tau) \right\}.$$

In analogy with (2.19), a necessary condition for a maximum of f on R^n can be written as follows: if f attains a maximum on R^n at a point y, then

$$\max_{u \in R^n} \min_{t \in E(y)} \left(\frac{\partial g}{\partial x}(y, t) \right)(u) = 0. \tag{2.29}$$

Obviously, condition (2.29) is equivalent to the condition

$$\max_{\|u\| \leq 1} \min_{t \in E(y)} \left(\frac{\partial g}{\partial x}(y, t) \right)(u) = 0. \tag{2.30}$$

Suppose that

$$H(y) = \left\{ z \in R^n \mid z = \frac{\partial g}{\partial x}(y, t), \ t \in E(y) \right\},$$

where $L(y)$ is the closed convex hull of $H(y)$. One can easily see that $H(y)$ and, consequently, $L(y)$ are bounded. Since the minimum of a linear functional on a set and on the convex hull of that set are the same, we have

$$\min_{t \in E(y)} \left(\frac{\partial g}{\partial x}(y, t) \right)(u) = \min_{z \in H(y)} z(u) = \min_{z \in H(y)} u(z) =$$
$$= \min_{z \in L(y)} u(z) = \min_{z \in L(y)} z(u),$$

from which it follows that (2.30) is equivalent to the condition

$$\max_{\|u\| \leqslant 1} \min_{z \in L(y)} z(u) = 0. \tag{2.31}$$

Since the sets $S = \{u \mid \|u\| \leqslant 1\}$ and $L(y)$ are convex, closed, and bounded, we obtain from (2.31) by using the minimax theorem (see, for example, [33]) the result

$$0 = \max_{\|u\| \leqslant 1} \min_{z \in L(y)} z(u) = \min_{z \in L(y)} \max_{\|u\| \leqslant 1} z(u) = \min_{z \in L(y)} \|z\|.$$

The necessary (and if f is a concave functional, sufficient) condition for a maximum is that the closed convex hull of the set $H(y)$ include the point 0. In particular, if $H(y)$ consists of a finite number of points, 0 must be included in the polyhedron defined by these points. On the other hand, if the set $E(y)$ consists of a single point t (in which case $H(y)$ consists of a single point $z = \frac{\partial g}{\partial x}(y, t)\Big)$, the condition, analogous to (2.20),

$$\frac{\partial g}{\partial x}(y, t) = 0$$

is satisfied.

3. Investigation of certain problems in approximation theory.

The necessary conditions for an extremum that were given in the first two sections can be used successfully to study numerous problems that are important in practice. We shall illustrate the procedure for applying these conditions with certain problems in approximation theory. Let us look at the following four problems.

1°. Problem 1. Let X denote a normed space, let $x_1, \ldots x_n$ denote members of X, and let p denote a sublinear functional defined on X. Let $\varphi(\alpha_1, \ldots, \alpha_n)$ denote a convex differentiable function of n variables such that the set

$$A = \{\alpha = (\alpha_1, \ldots, \alpha_n) \in R^n \mid \varphi(\alpha) \leqslant 1\}$$

is bounded in R^n. For a given element $y \in X$, let us find — out of all "polynomials" of the form

$$Q = \sum_{k=1}^{n} \alpha_k x_k,$$

where $\alpha = (\alpha_1, \ldots, \alpha_n) \in A$—the polynomial

$$Q_0 = \sum_{k=1}^{n} \overset{0}{\alpha}_k x_k$$

such that

$$p\left(\sum_{k=1}^{n} \overset{0}{\alpha}_k x_k - y\right) = \min_{\alpha \in A} p\left(\sum_{k=1}^{n} \alpha_k x_k - y\right). \tag{3.1}$$

We note first of all that problem 1 has a solution. This is true because this problem coincides with the problem of minimizing the functional p on the set

$$\Omega = \left\{ z \in X \,|\, z = \sum_{k=1}^{n} \alpha_k x_k - y \,|\, \alpha = (\alpha_1, \ldots, \alpha_n) \in A \right\}$$

and always has a solution since Ω is compact. Let us assume that

$$\min_{z \in \Omega} p(z) \neq \min_{x \in X} p(x),$$

in other words, that, if

$$\min_{x \in X} p(x) = 0,$$

then there exists an element z such that $p(z) = 0$ does not belong to Ω. In this case, condition (2.16) is valid and the polynomial Q_0 will be sought if and only if there exists an $h \in U_p^{Q_0 - y}$ such that

$$h(Q_0 - y) = \min_{z \in \Omega} h(z),$$

or, what amounts to the same thing,

$$\sum_{k=1}^{n} \overset{0}{\alpha}_k h(x_k) = \min_{\alpha \in A} \sum_{k=1}^{n} \alpha_k h(x_k).$$

Thus, if we know h, to find $\alpha_0=(\overset{0}{\alpha}_1, \ldots, \overset{0}{\alpha}_n)$, we need to solve the problem of minimizing the linear functional $\tilde{h}=(h(x_1), \ldots, h(x_n))$, defined on n-dimensional space, R^n on the set A. Obviously, the linear function $\tilde{h}$ is differentiable and it coincides with its gradient at an arbitrary point h. Let us first suppose $\tilde{h}\neq 0$. By using Lyusternik's theorem (see subsection 13° of Sec. 2), we can easily show that one of the following two situations is possible:

a) $\frac{\partial\varphi}{\partial\alpha_1}(\alpha_0)=\frac{\partial\varphi}{\partial\alpha_2}(\alpha_0)=\ldots=\frac{\partial\varphi}{\partial\alpha_n}(\alpha_0)=0;$

b) there exists a $\mu\neq 0$ such that

$$\frac{\partial\varphi}{\partial\alpha_1}(\overset{0}{\alpha}_1, \ldots, \overset{0}{\alpha}_n)=\mu h(x_1),$$
$$\frac{\partial\varphi}{\partial\alpha_2}(\overset{0}{\alpha}_1, \ldots, \overset{0}{\alpha}_n)=\mu h(x_2),$$
$$\cdots\cdots\cdots\cdots$$
$$\frac{\partial\varphi}{\partial\alpha_n}(\overset{0}{\alpha}_1, \ldots, \overset{0}{\alpha}_n)=\mu h(x_n).$$

Now, let us consider the case when

$$h(x_1)=h(x_2)=\ldots=h(x_n)=0. \tag{3.2}$$

We set

$$\Omega'=\left\{z\in X \mid z=\sum_{k=1}^{n}\alpha_k x_k \mid \alpha=(\alpha_1, \ldots, \alpha_n)\in R^n\right\}.$$

By using (2.24), we can easily show that (3.2) is a necessary and sufficient condition for p to have a minimum on the set $\Omega'-y$, that is, boundedness of $\alpha\in A$ proves in the present case to be unessential. Thus, we have

THEOREM 3.1. *If the polynomial*

$$Q_0=\sum_{k=1}^{n}\overset{0}{\alpha}_k x_k$$

is a solution of problem 1, then there exists a functional $h\in U_p^{Q_0-y}$ *such that either* $h(x_1)=h(x_2)=\ldots=h(x_n)=0$ *(in which case boundedness of* $\varphi(\alpha_1, \ldots, \alpha_n)\leqslant 1$ *proves to be unessential) or the coefficients* $\overset{0}{\alpha}_k$ *satisfy the following system of equations:*

$$\frac{1}{h(x_1)}\frac{\partial\varphi}{\partial\alpha_1}(\alpha_1, \ldots, \alpha_n)=\ldots=\frac{1}{h(x_n)}\frac{\partial\varphi}{\partial\alpha_n}(\alpha_1, \ldots, \alpha_n).$$

Here, it is assumed that if $h(x_k)=0$ for some k, the corresponding equation is replaced with the equation

$$\frac{\partial\varphi}{\partial\alpha_k}(\alpha_1, \ldots, \alpha_n)=0.$$

2°. Problem 2. Out of all n-times-continuously differentiable functions $x(t)$ defined on a closed interval $[0, T]$ that satisfy the two conditions

a) $x^{(n)}(t)$ satisfies on $[0, T]$ a Lipschitz condition with Lipschitz constant L,

b) $x(0)=x'(0)=\ldots=x^{(n)}(0)=0$,

find the one that best approximates a given function $z\in L^p([0, T])$ in the sense of $L^p([0, T])$, where $1<p<\infty$.

Let us denote by Ω^1 the subset of $L^p([0, T])$ consisting of all functions satisfying conditions a) and b). We need to find $y\in\Omega^1$ such that

$$\|y-z\|_{L^p}=\min_{x\in\Omega^1}\|x-z\|_{L^p}.$$

Let x denote a member of Ω^1. Then, it follows from condition a) that $x^{(n)}$ is an absolutely continuous function of $[0, T]$. Therefore, the function $x^{(n)}(t)$ has almost everywhere on $[0, T]$ a summable derivative, which we denote by $u(t)$. It also follows from condition a) that

$$\|u\|_{L^\infty}=\operatorname{vrai}\max_{t\in[0, T]}|u(t)|\leqslant L. \tag{3.3}$$

This is true because a Lipschitz condition with constant L is satisfied for $x^{(n)}$ and, hence, for arbitrary $t_0\in[0, T]$ and Δt_0 such that $(t_0+\Delta t_0)\in[0, T]$,

$$|x^{(n)}(t_0)-x^{(n)}(t_0+\Delta t_0)|\leqslant L|\Delta t_0|,$$

from which (3.3) follows. From the definition of $u(t)$ and condition b), it follows that

$$x^{(n)}(t)=\int_0^t u(t_1)\,dt_1;\; x^{(n-1)}(t)=\int_0^t\left(\int_0^{t_1} u(t_2)\,dt_2\right)dt_1,$$

$$\cdots\cdots\cdots\cdots\cdots\cdots$$

$$x(t)=\int_0^t\int_0^{t_1}\ldots\int_0^{t_n} u(t_{n+1})\,dt_{n+1}dt_n\ldots dt_1.$$

From the familiar formula in analysis,

$$x(t)=\frac{1}{n!}\int_0^t (t-\tau)^n u(\tau)\,d\tau. \tag{3.4}$$

Thus, every element x of Ω' can be represented in the form (3.4), where u satisfies the restriction (3.3). One can easily show that the converse is also true: if u satisfies (3.3), then the element x defined for a given u by means of formula (3.4) belongs to Ω'.

Let us look at the linear integral operator B that maps L^p into itself according to the formula

$$Bu(t)=\frac{1}{n!}\int_0^t (t-\tau)^n u(\tau)\,d\tau. \tag{3.5}$$

Let U_L denote the subset L^p consisting of all functions that are almost everywhere bounded on $[0, T]$ and that satisfy (3.3). As was shown above,

$$\Omega^1=\{x\in L^p([0, T])\,|\,x=Bu,\ u\in U_L\}.$$

Let us set $\Omega=\Omega^1-z$. Obviously, problem 2 consists in finding the element with the smallest norm on Ω. It is well known that the operator B defined by formula (3.5) is completely continuous (see, for example, [32, p. 277]). Furthermore, since U_L is weakly compact in L^p, the set Ω^1 and, consequently, Ω is compact. It follows, in particular, that problem 2 has a solution. Applying the necessary and sufficient condition for a minimum (2.18), we find that the element y is a solution of our problem if and only if there exists an $h\in L^q$, where $\frac{1}{p}+\frac{1}{q}=1$, such that

$$\|h\|_q = 1, \quad h(y - z) = \|y - z\|_p, \tag{3.6}$$
$$h(y) = \min_{x \in \Omega^1} h(x). \tag{3.7}$$

By using example 5) of subsection 4° of Sec. 3 of Chapter 1, we see that (3.6) is equivalent to the condition

$$h = \operatorname{sign}(y - z)(|y - z| / \|y - z\|_p)^{p-1}. \tag{3.8}$$

It follows from condition (3.7) that

$$y \in G_{\Omega^1} h, \tag{3.9}$$

where the mapping G_{Ω^1} is defined by formula (I.5.1). From the definition of Ω^1 and formula (I.5.3), we obtain

$$G_{\Omega^1} h = B(G_{U_L} B^* h) = \{x \in \Omega^1 \mid x = Bu, \; u \in G_{U_L}(B^* h)\}. \tag{3.10}$$

It is easy to describe the set $G_{U_L}(B^*h)$ by using (I.5.2) and (I.5.6). Suppose that

$$\begin{aligned} E_1 &= \{\tau \in [0, T] \mid B^*h(\tau) \neq 0\}, \\ E_2 &= \{\tau \in [0, T] \mid B^*h(\tau) = 0\}. \end{aligned} \tag{3.11}$$

Then, the general form of the elements $u(\tau) \in G_{U_L}(B^*h)$ is given by the formula

$$u(\tau) = \begin{cases} -L \operatorname{sign}(B^*h)(\tau), & \tau \in E_1, \\ v(\tau), & \tau \in E_2, \end{cases} \tag{3.12}$$

where $v(\tau)$ is a measurable function defined on E_2 and satisfying only the condition $|v(\tau)| \leqslant L$. Let us now calculate the functional B^*h. If we set

$$l(t, \tau) = \begin{cases} 0, & \tau > t, \\ 1, & \tau \leqslant t \end{cases}$$

and use the definition of a dual operator and formula (3.5), we have, for arbitrary $u \in L^p([0, T])$,

$$(B^*h)(u)=h(Bu)=\frac{1}{n!}\int_0^T h(t)\left(\int_0^T (t-\tau)^n u(\tau)\,d\tau\right)dt=$$

$$=\frac{1}{n!}\int_0^T h(t)\left(\int_0^T l(t,\tau)(t-\tau)^n u(\tau)\,d\tau\right)dt=\frac{1}{n!}\int_0^T u(\tau)\left(\int_0^T l(t,\tau)\times\right.$$

$$\left.\times(t-\tau)^n h(t)\,dt\right)d\tau=\frac{1}{n!}\int_0^T\left(\int_\tau^T (t-\tau)^n h(t)\,dt\right)u(\tau)\,d\tau,$$

from which it follows that

$$(B^*h)(\tau)=\frac{1}{n!}\int_0^T (t-\tau)^n h(t)\,dt. \tag{3.13}$$

By using the relations (3.8) - (3.13), we obtain the following necessary and sufficient conditions that the element y must satisfy: suppose that

$$E_1=\left\{\tau\in[0,\ T]\,\Big/\int_\tau^T (t-\tau)^n \operatorname{sign}(y(t)-z(t))\times\right.$$

$$\left.\times|y(t)-z(t)|^{p-1}\,dt\neq 0\right\};\quad E_2=[0,\ T]\diagdown E_1. \tag{3.14}$$

There exists a measurable function $v_0(\tau)$ defined on E_2 and satisfying the condition $|v(\tau)|\leqslant L$ such that

$$y(t)=-\frac{L}{n!}\int_{E_1(t)} (t-\tau)^n\times$$

$$\times\operatorname{sign}\left[\int_\tau^T (t-\tau)^n \operatorname{sign}(y(t)-z(t))\,|y(t)-z(t)|^{p-1}\,dt\right]d\tau+$$

$$+\frac{1}{n!}\int_{E_2(t)} (t-\tau)^n v_0(\tau)\,d\tau, \tag{3.15}$$

where

$$E_1(t)=E_1\cap[0,\ t]$$

and

$$E_2(t) = [0,\ T] \setminus E_1(t).$$

We note that, if $x \in \Omega^1$, then, for $t \in [0,\ T]$,

$$|x(t)| \leqslant \frac{1}{n!} \int_0^t |(t-\tau)^n u(\tau)|\, d\tau \leqslant \frac{L}{n!} \int_0^t (t-\tau)^n\, d\tau = \frac{L}{(n+1)!}\, t^{n+1}.$$

If the function z in question is such that, for all $t \in [0,\ T]$,

$$z(t) > \frac{1}{(n+1)!}\, L t^{n+1},$$

then it follows from (3.14) that

$$\mu(E_1) = \mu([0,\ T]) = T, \quad \mu(E_2) = 0.$$

But since sign $(y(t) - z(t)) = -1$ almost everywhere, we obtain in this case, by using (3.15),

$$y(t) = \frac{-1}{n!} L \int_0^t (t-\tau)^n (-1)\, d\tau = \frac{1}{(n+1)!}\, L t^{n+1}.$$

3°. Problem 3. In the space $L^p([0,\ T])$, where $1 < p < \infty$, let us consider the set

$$\Omega = \{x \in L^p \mid x \geqslant 0,\ \|x\| \leqslant 1\}.$$

Let f denote a differentiable convex functional defined on L^p and let z denote a member of L^p. Exhibit necessary and sufficient conditions that an element of y of Ω must satisfy if

$$f(y-z) = \min_{x \in \Omega} f(x-z). \tag{3.16}$$

Obviously, the functional f_z attains its minimum on Ω at the point y [for $x \in L^p$, we have $f_z(x) = f(x-z)$]. One can easily show that f_z is a differentiable functional. If, as usual, we denote the gradient of the functional f by F, then the gradient F_z of the functional f_z is calculated from the formula

$$F_z x = F(x-z). \tag{3.17}$$

It is also obvious that f_z is a convex functional. Since Ω is weakly compact, f_z attains a minimum on Ω; that is, our problem has a solution.

In this case, let us apply the necessary and sufficient condition for a minimum (2.12). On the basis of this, we get

$$F_z y(y) = \min_{x \in \Omega} F_z y(x).$$

If we set $(F_z y)^-(t) = \min((F_z y)(t), 0)$, we have[1] by using the results of example 4) of subsection 3° of Sec. 6 of Chapter 1,

$$\min_{x \in \Omega} F_z y(x) = \min_{x \in \Omega} \int_0^T (F_z y)(t)\, x(t)\, dt =$$

$$= \int_0^T (F_z y)^-(t) \left(\frac{|(F_z y)^-(t)|}{\|(F_z y)^-\|} \right)^{q-1} dt =$$

$$= -\int_0^T \|(F_z y)^-\|^{1-q} \, |(F_x y)^-(t)|^q \, dt = -\|(F_z y)^-\|.$$

Keeping (3.17) in mind, we obtain a necessary and sufficient condition for a minimum in the case of the present problem: the element y satisfies (3.16) if and only if

$$F(y-z)(y) = -\|(F(y-z))^-\|.$$

4°. Problem 4. In the space $C([0, T])$, let us consider the set

$$\Omega = \left\{ x \in C \,\middle|\, \max_{t \in [0, T]} x(t) = c \right\}.$$

Let f denote a differentiable functional defined on C and let z denote a member of C. Find necessary conditions that an element y (if it exists) must satisfy in order for us to have

$$f(y-z) = \min_{x \in \Omega} f(x-z). \tag{3.18}$$

[1]Here, $(F_z y)$ denotes an element of the space $L^q(E)$, where $\left(\frac{1}{p} + \frac{1}{q} = 1\right)$, that is generated by the functional $F_z y$.

We shall assume that the gradient of the functional f, that is, the operator F, is defined from C into L^1; that is, for arbitrary $x, u \in C$,

$$Fx(u) = \int_0^T (Fx)(t)\, u(t)\, dt,$$

where the function $(Fx)(t)$ is summable[1] on $[0, T]$. We note that the set Ω in the present case is not convex and is not weakly compact.

To investigate problem 4, we apply the necessary condition (2.2). It follows from this condition that the equation

$$\min_{u \in \overline{K_y(\Omega)}} F_z y(u) = 0 \tag{3.19}$$

must be satisfied at the point y being sought. Here, F_z denotes the gradient of the functional f_z, where $f_z(x) = f(x - z)$. In subsection 4° of Sec. 1, we showed that, in the present case, $\overline{K_y(\Omega)} = M_y$. Let us set

$$E_1(y) = \left\{ t \in [0, T],\ y(t) = \max_{\tau \in [0, T]} y(\tau) \right\},$$

and

$$E_2(y) = [0, T] \setminus E_1(y).$$

Then, according to Theorem (1.2),

$$M_y = \left\{ u \in C \,\middle|\, \max_{t \in E_1(y)} u(t) = 0 \right\}.$$

It follows from (3.19) that

$$\min_{u \in M_y} \int_0^T (F_z(y)(t)\, u(t))\, dt = \min_{u \in M_y} \left(\int_{E_1(y)} (F_z y)(t)\, u(t)\, dt + \right.$$
$$\left. + \int_{E_2(y)} (F_z y)(t)\, u(t)\, dt \right) = 0. \tag{3.20}$$

[1]Here, (Fx) denotes the function generating the functional Fx.

Consider the set $U=\{u\in C\,|\,u(t)=0,\ t\in E_1(y)\}$. Obviously, $U\subset M_y$. It follows from (3.20) that

$$\min_{u\in U}\int_{E_2(y)} F_z y(t)\,u(t)\,dt=0,$$

from which it follows that $(F_z y)(t)=0$, where $t\in E_2(y)$. Since $u(t)\leqslant 0$ for $u\in M_y$ and $t\in E_1(y)$, we obtain by using (3.20) the result that $(E_z y)(t)\leqslant 0$ for $t\in E_1(y)$. Using the fact that $F_z y=F(y-z)$, we can formulate the desired necessary condition for a minimum in the following manner: if a point y is a solution of problem 4, then

$$\begin{aligned}(F(y-z))(t)\leqslant 0 \quad (t\in E_1(y)),\\ (F(y-z))(t)=0 \quad (t\in E_2(y)).\end{aligned} \tag{3.21}$$

Let us consider, in particular, the case in which

$$f(x)=\int_0^T g(x(t))\,dt,$$

where $g(\alpha)$ is a continuously differentiable function defined for all real α the derivative $g'(\alpha)$ of which has the following properties: $g'(\alpha)<0$ for $\alpha<0$, $g'(0)=0$, and $g'(\alpha)>0$ for $\alpha>0$. (We note that, in this case, it is also easy to investigate the posed problem directly.) One can easily show, that under our assumptions, the functional f is differentiable and that for arbitrary $x,\ u\in C$,

$$\begin{aligned}Fx(u)=\lim_{\alpha\to 0}\int_0^T \frac{1}{\alpha}(g(x(t)+\alpha u(t))-g(x(t)))\,dt=\\ =\int_0^T g'(x(t))\,u(t)\,dt,\end{aligned}$$

from which it follows that $(Fx)(t)=g'(x(t))$. In the present case, condition (3.21) can be rewritten as follows:

$$\begin{aligned}g'(y(t)-z(t))\leqslant 0 \quad (t\in E_1(y)),\\ g'(y(t)-z(t))=0 \quad (t\in E_2(y)).\end{aligned}$$

By virtue of the assumptions regarding g' and the fact that $y(t)=C$ for $t\in E_1(y)$, these last inequalities are equivalent to the two relations

$$z(t) \geqslant C \quad (t \in E_1(y)), \tag{3.22a}$$

$$y(z) = z(t) \quad (t \in E_2(y)). \tag{3.22b}$$

Let us now consider separately the following two cases:

1) $\max\limits_{t \in [0, T]} z(t) \geqslant C$. Let us set

$$\tilde{E}_1(z) = \{t \in [0, T] \mid z(t) \geqslant C\},$$

and

$$\tilde{E}_2(z) = [0, T] \setminus \tilde{E}_1(z).$$

Define

$$y(t) = \begin{cases} c, & t \in E_1(z), \\ z(t), & t \in E_2(z). \end{cases}$$

One can easily show that $y(t)$ is a continuous function and that it belongs to Ω. Obviously $E_1(y) = \tilde{E}_1(z)$ and $E_2(y) = \tilde{E}_2(z)$. Therefore, it follows from (3.22a) and (3.22b) that y satisfies our necessary condition. It follows from the assumption on g' that $g(\alpha)$ is a decreasing function for $\alpha \leqslant 0$ and an increasing one for $\alpha \geqslant 0$. By using this fact, we can easily show that the function y does indeed approximate z "in the sense f" in the best possible way and hence is a solution of our problem.

2) $\max\limits_{t \in [0, T]} z(t) < C$. In this case, it follows from (3.22a) that $E_1(y) = \varnothing$. Therefore, if the problem had a solution, the best approximation would be given by the function $y(t) = z(t)$ for $(t \in [0, T])$. However, since z is not a member of Ω, the problem does not have a solution in this case.

CHAPTER 3

Successive Approximation Methods

1. The conditional-gradient method.

1°. Let f denote a functional defined on a Banach space X. Suppose that f is differentiable with respect to directions and that it attains a minimum on X. To solve the problem of minimizing f on X, the method of fastest descent proves to be extremely convenient. The outline of this method is described, for example, in [32, Chapter XV]. This procedure consists in the following:

Let x denote a member of X. Let us calculate the derivative of the functional f at the point x with respect to the direction u:

$$\frac{1}{\|u\|} \lim_{\alpha \to +0} \frac{f(x+\alpha u) - f(x)}{\alpha} = \frac{f'_x(u)}{\|u\|} = f'_x\left(\frac{u}{\|u\|}\right),$$

and let us choose an element u_0 (under the assumption that it exists) minimizing this derivative. A descent is made in the direction u_0 (the direction of the antigradient). It follows from the relation written above that, to find the antigradient, we need to know how to solve the problem of minimizing the functional f'_x on a unit sphere in the space X. In particular, if f is a differentiable functional (and if f'_x is a linear functional), this problem reduces to that of minimizing a linear functional on a unit sphere (or, what amounts to the same thing, on a unit ball) in the space X. Once we have chosen the *direction* of the descent u_0, we determine the *amount* of descent. To do this, we consider, for $\alpha \in [0, \infty)$, the function

$$g_{x, u_0}(\alpha)=f(x+\alpha u_0).$$

Assuming that $(g_{x, u_0})'_{\text{r.h.}}(0)=f'_x(u_0)<0$* and that the derivative $(g_{x, u_0})'(\alpha)$ exists and is continuous for $\alpha>0$, we can guarantee that the function $g_{x, u_0}(\alpha)$ decreases on the interval $[0, \alpha_0]$, where α_0 is the smallest positive root of the equation

$$(g'_{x, u_0})(\alpha)=0.$$

If the point x was the nth approximation of the minimum, we take as the $(n+1)$st approximation the element $x'=x+\alpha_0 u_0$.

The conditional-gradient method, which we shall examine in the present section, is a generalization of the method of fastest descent for solving the problem of minimizing a differentiable functional f on a convex weakly compact set Ω. The direction of descent in this method is sought by minimizing the linear functional Fx on the set Ω. (As before, we denote by F the gradient of the functional f.) The amount of descent can be chosen on the basis of different considerations, to some extent the same considerations as in the method of fastest descent.

We note that throughout this section, we shall assume the solution of the problem of minimizing a linear functional on Ω to be known. The method of the conditional gradient has been applied by various authors for solving specific problems [13, 14, 57]. In its general form, it is examined in [22, 23, 39, 54).

2°. Let Ω denote a convex weakly compact subset of X, let f denote a differentiable functional defined on X, and let F denote the operator representing the gradient functional f. The method in question is described as follows: we choose and fix a number a in $[0, 1]$.

1) As our first approximation, let us take an arbitrary element $x_1 \in \Omega$.

2) Suppose that elements x_n are already chosen. If $x_n \in G_\Omega F x_n$, where x_n is such that

*Ed. Note: Subscripts r.h. denote right-handed.

$$Fx_n(x_n) = \min_{x \in \Omega} Fx_n(x)),$$

the process terminates. Otherwise, let us find $\bar{x}_n \in G_\Omega Fx_n$. Then, for $\alpha \in [0, a_n]$, where a_n is an arbitrary number satisfying the inequalities

$$a \leqslant a_n \leqslant 1, \qquad (1.1)$$

let us consider the function

$$g_n(\alpha) = f(x_n + \alpha(\bar{x}_n - x_n)).$$

Suppose that $g_n(\alpha)$ attains a minimum on $[0, a_n]$ at a point α_n. Since

$$g'_n(0) = Fx_n(\bar{x}_n - x_n) = \min_{x \in \Omega} Fx_n(x) - Fx_n(x_n) < 0,$$

we know that $\alpha_n \neq 0$. As our $(n+1)$st approximation, we take the element

$$x_{n+1} = x_n + \alpha_n(\bar{x}_n - x_n).$$

We have now constructed sequences (finite or infinite)

$$\begin{aligned} &x_1, x_2, \ldots, x_n, \ldots, \\ &\bar{x}_1, \bar{x}_2, \ldots, \bar{x}_n, \ldots. \end{aligned} \qquad (1.2)$$

It follows from the definition of x_{n+1} that

$$f(x_{n+1}) < f(x_n) \quad (n = 1, 2, \ldots).$$

Suppose that $\bar{x} \in G_\Omega Fx$, for $x \in \Omega$. Let us consider a functional φ defined on Ω by

$$\varphi(x) = Fx(x - \bar{x})$$

or, what amounts to the same thing,

$$\varphi(x) = Fx(x) - \min_{z \in \Omega} Fx(z)).$$

It follows from the definition of $\bar{x}$ that $\varphi(x) \geqslant 0$ for $x \in \Omega$. Below, we shall show that, under certain natural assumptions regarding f,

$$\inf_{x \in \Omega} \varphi(x) = \inf_{x \in \Omega} Fx(x - \bar{x}) = 0, \tag{1.3}$$

and the sequence (1.2) [if it is infinite] is a minimizing sequence for the functional φ on Ω. If (1.2) has n terms, then the element $x_n = \inf_{x \in \Omega} \varphi(x)$.

Let us suppose that the infimum in (1.3) is attained, that is, that there exists an element $y \in \Omega$ such that

$$Fy(y - \bar{y}) = 0,$$

or, what amounts to the same thing,

$$Fy(y) = \min_{x \in \Omega} Fy(x). \tag{1.4}$$

The necessary condition for a minimum (II.2.12) is satisfied at the point y. (If f is a convex functional, this condition is also sufficient.) We shall refer to points at which (1.4) is satisfied as stationary points of the functional f on the set Ω. Since the sequence (1.2) is (if it is infinite) a minimizing sequence for the functional φ and since elements on which the minimum φ is attained are stationary points of f, the cluster points (in the weak sense) of the sequence (1.2) are stationary points of f on Ω. (If f is a convex functional, they are the minima of f on Ω.) If (1.2) has n terms, then x_n is a stationary point of f on Ω. Thus, the method of the conditional gradient enables us to find only stationary points. We note that if a stationary point y belongs to the interior of Ω, then, on the basis of (1.4), we have $Fy = 0$. The relation $Fy = 0$ is a necessary condition for a local extremum (maximum or minimum) of the functional f on all X [see (II.2.20)]. Thus, in the case in question, a (local) maximum f can in general be attained at an element y. Speaking somewhat informally, we may say that, in this case, the stationarity of y is unconnected with Ω. On the other hand, if the point y is stationary but $Fy(x - y) \not\equiv 0$ on Ω, then, as one can easily show, a (local) maximum of f on X cannot be attained at the point y. In this case, the stationarity of y is connected with Ω.

3°. Let us look at the question of the convergence of the method. Here, we assume without further mention that the sequence (1.2) is infinite. Let us first prove the following lemmas:

LEMMA 1.1. *Suppose that an operator F, the gradient of a functional f, satisfies a Lipschitz condition for x', $x'' \in \Omega$ on a set Ω:*

$$\|Fx' - Fx''\| \leqslant L\|x' - x''\|. \tag{1.5}$$

Then, F is continuous on Ω and, in addition,

$$\sup_{x \in \Omega} \|Fx\| < \infty, \quad \sup_{x \in \Omega} |f(x)| < \infty.$$

Proof: The continuity of F follows immediately from (1.5). Let x_0 denote a member of Ω. Then, for arbitrary $x \in \Omega$, we have

$$\|Fx\| \leqslant \|Fx_0\| + \|Fx_0 - Fx\| \leqslant \|Fx_0\| + L\|x_0 - x\| \leqslant \|Fx_0\| + LD,$$

where D denotes the diameter of Ω:

$$D = \sup_{x', x'' \in \Omega} \|x' - x''\|.$$

(We note that weak compactness of Ω implies that Ω is bounded, so that $D < \infty$.) Thus,

$$\sup_{x \in \Omega} \|Fx\| = M < \infty.$$

Similarly, by using Lagrange's formula [see (I.3.21)], we have, for $x \in \Omega$,

$$\begin{aligned} |f(x)| \leqslant |f(x_0)| + |f(x_0) - f(x)| = |f(x_0)| + \\ + |F(x' + \theta(x' - x''))(x' - x'')| \leqslant |f(x_0)| + \\ + \|F(x' + \theta(x' - x''))\| \cdot \|x' - x''\| \leqslant |f(x_0)| + MD. \end{aligned}$$

Here, $\theta \in [0, 1]$. This proves that $\sup_{x \in \Omega} |f(x)| < \infty$.

LEMMA 1.2. *Suppose that the operator F, the gradient of a functional f, satisfies the Lipschitz condition (1.5) on a set Ω. Let x' and x'' denote members of Ω and define $x_\alpha = x' + \alpha(x'' - x')$. Then,*

$$f(x_\alpha) \leqslant f(x') + \alpha Fx'(x'' - x') + \frac{\alpha^2}{2} L\|x'' - x\|^2. \tag{1.6}$$

Proof: In view of formula (I.3.23) and the properties of an integral of an abstract function (see [40, Chapter 8, Sec. 1]), we have

$$f(x_\alpha)=f(x')+\int_0^1 F(x'+\tau(x_\alpha-x'))(x_\alpha-x')\,d\tau=f(x')+$$

$$+Fx'(x_\alpha-x')-Fx'(x_\alpha-x')+\int_0^1 F(x'+\tau(x_\alpha-x'))(x_\alpha-x')\,d\tau=$$

$$=f(x')+Fx'(x_\alpha-x')+\int_0^1 [F(x'+\tau(x_\alpha-x'))-Fx'](x_\alpha-x')\,d\tau\leqslant$$

$$\leqslant f(x')+Fx'(x_\alpha-x')+\int_0^1 \|F(x'+\tau(x_\alpha-x'))-Fx'\|\cdot\|x_\alpha-x'\|\,d\tau\leqslant$$

$$\leqslant f(x')+Fx'(x_\alpha-x')+\int_0^1 L\tau\|x_\alpha-x'\|^2\,d\tau=$$

$$=f(x')+Fx'(x_\alpha-x')+\frac{L}{2}\|x_\alpha-x'\|^2=$$

$$=f(x')+\alpha Fx'(x''-x')+\frac{\alpha^2}{2}L\|x''-x'\|^2,$$

which completes the proof.

THEOREM 1.1. *Suppose that an operator F satisfies the Lipschitz condition (1.5) on a set Ω. Then, the sequence (1.2) has the property that*

$$\lim_{n\to\infty} Fx_n(x_n-\bar{x}_n)=0.$$

Proof: At the nth step, the amount of descent α_n was chosen in such a way that $f(x_{n+1})\leqslant f(x_n+\alpha(\bar{x}_n-x_n))$ for $\alpha\in(0, a_n]$. Since $\alpha\in(0, a_n]$ by virtue of (1.1), this last inequality is *a fortiori* is valid for $\alpha\in[0, a]$. Keeping this fact in mind and using formula (1.6), we see that, for all n and all $\alpha\in[0, a]$,

$$f(x_{n+1})\leqslant f(x_n)+\alpha Fx_n(\bar{x}_n-x_n)+\frac{\alpha^2}{2}L\|\bar{x}_n-x_n\|^2\leqslant$$
$$\leqslant f(x_n)+\alpha Fx_n(\bar{x}_n-x_n)+\alpha^2\frac{LD^2}{2}, \quad (1.7)$$

where, as before, D is the diameter of Ω. Since

$$Fx_n(x_n-\bar{x}_n)\geqslant 0,$$

by virtue of the definition of the element $\bar{x}_n$, we obtain by using (1.7)

$$0 \leqslant Fx_n(x_n - \bar{x}_n) \leqslant \frac{f(x_n) - f(x_{n+1})}{\alpha} + \alpha \frac{LD^2}{2}.$$

If we take $\alpha_0 \in (0, \varepsilon]$, where ε is an arbitrary positive number, we can find a number N such that, for $n > N$,

$$\frac{f(x_n) - f(x_{n+1})}{\alpha_0} < \varepsilon.$$

(This is possible because the sequence $\{f(x_n)\}$ decreases monotonically and, by virtue of the boundedness of f on Ω, converges to a finite limit.) For $n \geqslant N$, we have

$$0 \leqslant Fx_n(x_n - \bar{x}_n) < \varepsilon\left(1 + \frac{LD^2}{2}\right),$$

which completes the proof of the theorem.

Remark. If Ω is compact, it follows from this theorem that the cluster points of the sequence $\{x_n\}$ are stationary points of f on Ω. As will be shown below, this assertion remains valid if we require only continuity of the operator F.

THEOREM 1.2. *Suppose that Ω is compact and that the operator F is continuous. Then, the cluster points of the sequence (1.2) are stationary points of f on Ω.*

Proof: Let us choose a sequence of numbers $n_1, n_2, \ldots, n_i, \ldots$ such that $\lim x_{n_i} = y$ and $\lim \bar{x}_{n_i} = \bar{y}$. From the formula on finite increment for $\alpha \in [0, a]$, we have

$$f(x_{n_i} + \alpha(\bar{x}_{n_i} - x_{n_i})) = f(x_{n_i}) + \alpha Fx_{n_i}(\bar{x}_{n_i} - x_{n_i}) + \alpha\psi_{n_i}(\alpha),$$

where

$$\psi_{n_i}(\alpha) \xrightarrow[\alpha \to +0]{} 0.$$

We note that, under the conditions of the theorem, the functional f is continuous. Therefore, by taking these limits, we have

$$f(y + \alpha(\bar{y} - y)) = f(y) + \alpha Fy(\bar{y} - y) + \alpha\psi(\alpha), \tag{1.8}$$

where

$$\psi(\alpha)=\lim_{i\to\infty}\psi_{n_i}(\alpha).$$

Let us define $g(\alpha)=f(y+\alpha(\bar{y}-y))$. In this notation, (1.8) becomes

$$g(\alpha)=g(0)+\alpha g'(0)+\alpha\psi(\alpha).$$

It now follows from the definition of a derivative that

$$\lim_{\alpha\to+0}\psi(\alpha)=0.$$

Remembering that

$$f(x_{n_{i+1}})\leqslant f(x_{n_i}+\alpha(\bar{x}_{n_i}-x_{n_i}))=$$
$$=f(x_{n_i})+\alpha Fx_{n_i}(\bar{x}_{n_i}-x_{n_i})+\alpha\psi_{n_i}(\alpha)$$

and

$$Fx_{n_i}(x_{n_i}-\bar{x}_{n_i})\geqslant 0,$$

we obtain

$$0\leqslant Fx_{n_i}(x_{n_i}-\bar{x}_{n_i})\leqslant\frac{f(x_{n_{i+1}})-f(x_{n_i})}{\alpha}+\psi_{n_i}(\alpha).$$

If we fix $\alpha\in(0,\ a]$ and take the limit as $i\to\infty$, we obtain

$$0\leqslant Fy(y-\bar{y})\leqslant\psi(\alpha).$$

It is easy to show, by taking the limit in the inequality

$$Fx_{n_i}(\bar{x}_{n_i})\leqslant Fx_{n_i}(\Omega),$$

that $\bar{y}\in G_\Omega Fy$. The assertion of the theorem now follows from the fact that

$$\lim_{\alpha\to+0}\psi(\alpha)=0.$$

This completes the proof of the theorem.

Remark: One can easily see that, under the conditions of the theorem,

$$\lim_{n \to \infty} Fx_n (x_n - \overline{x}_n) = 0.$$

Suppose that X is a reflexive space. We shall say that an operator F is strongly continuous in the sphere $S_r(0)$ if it maps every weakly convergent sequence of elements of that sphere into a sequence that converges in the norm. We know (see, for example, Theorems 1.2 and 8.2 of [6]) that strong continuity of the gradient in a sphere implies weak continuity of the functional in that sphere. By using this fact and reasoning as in the proof of the preceding theorem, one can easily verify the validity of the following assertion:

THEOREM 1.3. *If F is strongly continuous in a sphere $S_r(0)$ containing Ω, then the cluster points (in the weak sense) of the sequence (1.2) are stationary.*

Let us suppose now that f is a convex functional. Then, f is weakly continuous from below and hence attains a minimum on Ω. Define $Q = \min_{x \in \Omega} f(x)$.

THEOREM 1.4. *Suppose that the conditions of any one of the Theorems 1.1 - 1.3 are satisfied, and that, in addition, f is a convex functional. Then,*

$$\lim_{n} f(x_n) = Q.$$

Proof: By applying Lagrange's formula and keeping the convexity of f in mind [see (I.3.22)], we have, for arbitrary $x \in \Omega$,

$$f(x) - f(x_n) = F(x_n + \theta(x - x_n))(x - x_n) \geqslant Fx_n (x - x_n),$$

from which we get

$$\min_{x \in \Omega} (f(x) - f(x_n)) = Q - f(x_n) \geqslant \min_{x \in \Omega} Fx_n (x - x_n) = Fx_n (\overline{x}_n - x_n).$$

It follows from this that

$$0 \leqslant f(x_n) - Q \leqslant Fx_n (x_n - \overline{x}_n). \tag{1.9}$$

The conclusion of the theorem then follows.

Remark 1. Theorem 1.4 shows that, in the case of a convex functional, the sequence (1.2) is a minimizing sequence. In particular, the minimum being sought is attained at the cluster points (in the weak sense) of that sequence. If the functional is strongly convex, the sequence (1.2) converges weakly to a unique point representing a minimum. If, in addition, f satisfies condition (I.4.1), it follows from Theorem 4.2 of Chapter 1 that the sequence x_n converges to the point y in norm. This sequence also converges in norm when the set Ω satisfies condition (I.4.4) and $\inf_{x \in \Omega} \|Fx\| = \varepsilon > 0$ (see remark following Theorem 4.3 of Chapter 1).

Remark 2. Inequalities (1.9) provide a convenient *a posteriori* estimate of the convergence.

4°. In solving many problems, the determination of the exact value of the direction of descent (of the element $\bar{x}_n$) and the magnitude of the descent (the number α_n) is complicated. In connection with this, it is of interest to determine a method in which $\bar{x}_n$ and α_n are determined approximately and to see whether it is possible to guarantee convergence of that method.

The method that we shall consider is described as follows: We choose and fix numbers $a \in (0, 1]$, $s \in [1, \infty)$, and $\lambda \in (0, 1)$. For x_1 let us take an arbitrary element of Ω. Suppose that x_n is already defined and

$$\min_{x \in \Omega} Fx_n(x - x_n) = -u_n.$$

If $u_n = 0$, the process terminates. If $u_n > 0$, we find an element $\bar{\bar{x}}_n$ satisfying the condition

$$-u_n \leqslant Fx_n(\bar{\bar{x}}_n - x_n) \leqslant -\frac{u_n}{s}.$$

Now, let a_n denote an arbitrary number in the interval $[a, 1]$. Let us set $f(x_n) = v_n$ and

$$\min_{\alpha \in [0, a_n]} f(x_n + \alpha(\bar{\bar{x}}_n - x_n)) = w_n.$$

Let us then find $\alpha_n' \in [0, a_n]$ such that

$$f(x_n + \alpha_n'(\overline{\overline{x}}_n - x_n)) \leqslant v_n + \lambda(w_n - v_n).$$

As the $(n+1)$st approximation, let us take the element

$$x_{n+1} = x_n + \alpha_n'(\overline{\overline{x}}_n - x_n).$$

Assuming that the sequence

$$x_1, x_2, \ldots, x_n, \ldots \tag{1.10}$$

thus constructed is infinite, let us prove

THEOREM 1.5. *Suppose that an operator F satisfies the Lipschitz condition (1.5). Then, the sequence (1.10) has the property that*

$$\lim_n Fx_n(x_n - \overline{x}_n) = 0.$$

The proof of the theorem rests on

LEMMA 1.3. *If an operator F satisfies the Lipschitz condition (1.5), then the sequence (1.10) is such that*

$$\lim_n Fx_n(\overline{\overline{x}}_n - x_n) = 0.$$

Proof: We note first of all that

$$f(x_{n+1}) = f(x_n + \alpha_n'(\overline{\overline{x}}_n - x_n)) \leqslant v_n + \lambda(w_n - v_n) =$$
$$= f(x_n) + \lambda\Big[\min_{\alpha \in [0, a_n]} f(x_n + \alpha(\overline{\overline{x}}_n - x_n)) - f(x_n)\Big] \leqslant f(x_n).$$

Since the functional f is, by virtue of Lemma 1.1, bounded on Ω, there exists

$$\lim_n f(x_n) = Q > -\infty.$$

Let us now suppose that the lemma is not true. Then, there exists a number $\rho > 0$ and a subsequence $\{x_{n_k}\}$ of the sequence $\{x_n\}$ such that

$$Fx_{n_k}(\overline{\overline{x}}_{n_k} - x_{n_k}) \leqslant -\rho < 0.$$

for every k. It follows from Lemma 1.2 that

$$f\left(x_{n_k}+\alpha\left(\bar{\bar{x}}_{n_k}-x_{n_k}\right)\right)\leqslant f\left(x_{n_k}\right)+\alpha F x_{n_k}\left(\bar{\bar{x}}_{n_k}-x_{n_k}\right)+ \\ +\frac{\alpha^2}{2}L\left\|\bar{\bar{x}}_{n_k}-x_{n_k}\right\|^2\leqslant f\left(x_{n_k}\right)+\alpha\left(-\rho+\alpha\frac{LD^2}{2}\right),$$

where D is the diameter of Ω. Obviously, there exists a sufficiently small number α_0 in $(0, a]$ that

$$-\rho+\alpha_0\frac{LD^2}{2}\leqslant-\frac{\rho}{2}.$$

Furthermore, for sufficiently large k,

$$f(x_{n_k})-Q\leqslant\alpha_0\frac{\rho}{4}.$$

For such values of k,

$$f\left(x_{n_k}+\alpha_0\left(\bar{\bar{x}}_{n_k}-x_{n_k}\right)\right)\leqslant f\left(x_{n_k}\right)+\alpha_0\left(-\rho+\alpha_0\frac{LD^2}{2}\right)\leqslant \\ \leqslant\frac{\alpha_0\rho}{4}+Q-\frac{\alpha_0\rho}{2}=Q-\frac{\alpha_0\rho}{4},$$

from which we get

$$w_{n_k}=\min_{\alpha\in[0,\,a]}f\left(x_{n_k}+\alpha\left(\bar{\bar{x}}_{n_k}-x_{n_k}\right)\right)\leqslant Q-\frac{\alpha_0\rho}{4}.$$

Thus,

$$f(x_{n_{k+1}})\leqslant v_{n_k}+\lambda\left(w_{n_k}-v_{n_k}\right)=\lambda w_{n_k}+(1-\lambda)\,v_{n_k}\leqslant \\ \leqslant\lambda\left(Q-\frac{\alpha_0\rho}{4}\right)+(1-\lambda)\,f\left(x_{n_k}\right).$$

Taking the limit as $k\to\infty$, we have

$$Q\leqslant Q-\lambda\alpha_0\rho\frac{1}{4},$$

which is impossible since $\lambda\alpha_0\rho>0$. This completes the proof of the lemma.

Proof of Theorem 1.5. Let us suppose that the theorem is not true. Then, there exists a subsequence $\{x_{n_k}\}$ of the sequence (1.10) such that

$$Fx_{n_k}\left(\bar{x}_{n_k} - x_{n_k}\right) \leqslant -\rho < 0$$

for every k. Remembering the rule by which the element $\bar{\bar{x}}_{n_k}$ is chosen, we obtain

$$Fx_{n_k}\left(\bar{\bar{x}}_{n_k} - x_{n_k}\right) \leqslant \frac{1}{s} Fx_{n_k}\left(\bar{x}_{n_k} - x_{n_k}\right) \leqslant -\frac{\rho}{s},$$

which contradicts Lemma 1.3.

5°. Suppose that a functional f is such that its gradient F satisfies a Lipschitz condition on a set Ω. Let us suppose that x' and x'' are elements of Ω such that $Fx'(x'' - x') < 0$ and that, for $\alpha \in [0, 1]$, equality holds in (1.6):

$$f(x' + \alpha(x'' - x')) = f(x') + \alpha Fx'(x'' - x') + \frac{\alpha^2}{2} L \| x'' - x' \|^2.$$

Then, as one can easily show, $g(\alpha) = f(x' + \alpha(x'' - x'))$ attains its minimum on the interval $[0, 1]$ either at the point

$$\alpha = \frac{-Fx'(x'' - x')}{L \| x'' - x' \|^2}$$

or at the point

$$\alpha = 1$$

according to

$$-\frac{Fx'(x'' - x')}{L \| x'' - x' \|^2} < 1 \quad \text{or} \quad -\frac{Fx'(x'' - x')}{L \| x'' - x' \|^2} \geqslant 1$$

respectively.

Suppose that the element x_n is chosen. Remembering that $a_n = 1$ [where a_n is the α_n appearing in (1.1)] and assuming that, for $\alpha \in [0, 1]$,

$$f(x_n + \alpha(\overline{x}_n - x_n)) \approx f(x_n) + \alpha F x_n(\overline{x}_n - x_n) + \frac{\alpha^2}{2} L \| \overline{x}_n - x_n \|^2,$$

we can choose the magnitude of the descent as follows:

$$\alpha_n = \min\left(1, \frac{F x_n (x_n - \overline{x}_n)}{L \| x_n - \overline{x}_n \|^2}\right).$$

It turns out that the method converges also when α_n is chosen from the formula

$$\alpha_n = \min\left(1, \gamma_n \frac{F x_n (x_n - \overline{x}_n)}{\| x_n - \overline{x}_n \|^2}\right), \tag{1.11}$$

where γ_n is any number satisfying the inequalities

$$\varepsilon_1 \leqslant \gamma_n \leqslant \frac{2 - \varepsilon_2}{L}$$

where $0 < \varepsilon_1 < \frac{2}{L}$ and $0 < \varepsilon_2 < 2 - L\varepsilon_1$, the ε's being independent of n. Let x_1 denote a member of Ω. Consider the sequence

$$x_1, x_2, \ldots, x_n, \ldots, \tag{1.12}$$

where $x_{n+1} = x_n + \alpha_n(\overline{x}_n - x_n)$ with α_n is chosen according to formula (1.11) for $n = 1, 2, \ldots$.

THEOREM 1.6. *The sequence (1.12) has the property that*

$$\lim_{n \to \infty} F x_n (x_n - \overline{x}_n) = 0.$$

Proof: Let us show first that the sequence $\{f(x_n)\}$ decreases monotonically. Suppose that

$$\delta_n = f(x_n) - f(x_{n+1}).$$

By using formula (1.6), we have

$$f(x_{n+1}) = f(x_n + (x_{n+1} - x_n)) \leqslant f(x_n) + \\ + F x_n (x_{n+1} - x_n) + \frac{L}{2} \| x_{n+1} - x_n \|^2,$$

from which we see that

$$-\delta_n = f(x_{n+1}) - f(x_n) \leqslant -\alpha_n F x_n (x_n - \bar{x}_n) + \frac{L\alpha_n^2}{2} \| x_n - \bar{x}_n \|^2. \tag{1.13}$$

Let us consider separately two cases:

1) $\alpha_n = 1$. In this case,

$$\gamma_n \frac{F x_n (x_n - \bar{x}_n)}{\| x_n - \bar{x}_n \|^2} \geqslant 1.$$

Therefore, by using (1.13), we have

$$\begin{aligned} -\delta_n &\leqslant -F x_n (x_n - \bar{x}_n) + \frac{L}{2} \| x_n - \bar{x}_n \|^2 = \\ &= F x_n (x_n - \bar{x}_n) \left(\frac{L}{2} \frac{\| x_n - \bar{x}_n \|^2}{F x_n (x_n - \bar{x}_n)} - 1 \right) \leqslant \\ &\leqslant F x_n (x_n - \bar{x}_n) \left(\frac{L\gamma_n}{2} - 1 \right) \leqslant -\frac{\varepsilon_2}{2} F x_n (x_n - \bar{x}_n). \end{aligned} \tag{1.14}$$

2) $\alpha_n = \gamma_n \dfrac{F x_n (x_n - \bar{x}_n)}{\| x_n - \bar{x}_n \|^2} \leqslant 1.$

From (1.13), we have

$$\begin{aligned} -\delta_n &\leqslant -\gamma_n \frac{[F x_n (x_n - \bar{x}_n)]^2}{\| x_n - \bar{x}_n \|^2} + \frac{L\gamma_n^2}{2} \frac{[F x_n (x_n - \bar{x}_n)]^2}{\| x_n - \bar{x}_n \|^2} = \\ &= \gamma_n \left(\frac{L\gamma_n}{2} - 1 \right) \frac{[F x_n (x_n - \bar{x}_n)]^2}{\| x_n - \bar{x}_n \|^2}. \end{aligned}$$

Remembering that $\gamma_n \leqslant \dfrac{2 - \varepsilon_2}{L}$, we obtain

$$\frac{L\gamma_n}{2} - 1 \leqslant \frac{L}{2} \frac{2 - \varepsilon_2}{L} - 1 = -\frac{\varepsilon_2}{2}.$$

Since $\gamma_n \geqslant \varepsilon_1$, we have

$$\gamma_n \left(\frac{L\gamma_n}{2} - 1 \right) \leqslant -\frac{\varepsilon_1 \varepsilon_2}{2}.$$

Thus, in the present case

$$-\delta_n \leqslant -\frac{\varepsilon_1 \varepsilon_2}{2} \frac{[F x_n (x_n - \bar{x}_n)]^2}{\| x_n - \bar{x}_n \|^2}. \tag{1.15}$$

In both cases, $\delta_n = f(x_n) - f(x_{n+1}) \geqslant 0$; that is, the sequence $\{f(x_n)\}$ decreases monotonically. On the basis of Lemma 1.1, the functional f is bounded on Ω, so that $\lim f(x_n)$ exists and is equal to zero. It follows, in particular, that $\delta_n \to 0$. Then, by virtue of (1.14) and (1.15), we obtain

$$0 \leqslant Fx_n(x_n - \bar{x}_n) \leqslant \max\left(\frac{2\delta_n}{\varepsilon_2},\ \nu\sqrt{\delta_n}\right), \tag{1.16}$$

where

$$\nu = \sqrt{\frac{2}{\varepsilon_1 \varepsilon_2}}\, D. \tag{1.17}$$

The assertion in the theorem follows from (1.16).

Remark: In the proof of the theorem, we showed in effect that, by choosing the coefficients α_n in accordance with formula (1.11), we decreased the value of the functional (because, if x_n is a stationary point, then $Fx_n(x_n - \bar{x}_n) > 0$ and hence $\delta_n > 0$). However, it is obvious that, by choosing α_n in the same way as in subsection 2° (making the fastest descent), we decrease the value of the functional to a greater degree. On the other hand, in the majority of cases, it is simpler to calculate α_n from formula (1.11) than to find the minimum of the function $g(\alpha) = f(x_n + \alpha(\bar{x}_n - x_n))$.

Reasoning as in the proof of Theorem 1.4, we can easily show that, if f is a convex functional, then the sequence (1.12) is a minimizing sequence. Inequalities (1.9), which give an *a posteriori* estimate of the convergence, remain valid in the present case. Let us give an estimate of the speed of convergence.

THEOREM 1.7. *If f is a convex functional and*

$$Q = \min_{x \in \Omega} f(x),$$

then the sequence (1.12) is such that

$$f(x_n) - Q = O\left(\frac{1}{n}\right).$$

The proof of the theorem rests on

LEMMA 1.4. *Let* $\{\lambda_n\}$ *denote a sequence of positive numbers such that*

$$\lambda_n - \lambda_{n+1} \geqslant q\lambda_n^2, \tag{1.18}$$

where q is a positive constant. Then,

$$\lambda_n = O\left(\frac{1}{n}\right).$$

Proof: Let $\lambda_n = \frac{C_n}{n}$. It will be sufficient to show that the sequence $\{C_n\}$ is bounded.

Using (1.18), we obtain

$$\lambda_n - \lambda_{n+1} = \frac{C_n}{n} - \frac{C_{n+1}}{n+1} = \frac{C_n}{n}\left(1 - \frac{C_{n+1}}{C_n}\frac{n}{n+1}\right) \geqslant q\frac{C_n^2}{n^2},$$

so that

$$1 - \frac{C_{n+1}}{C_n}\left(\frac{n}{n+1}\right) \geqslant q\frac{C_n}{n}.$$

Transforming this last inequality, we have

$$\left(1 + \frac{1}{n}\right)\left(1 - q\frac{C_n}{n}\right) \geqslant \frac{C_{n+1}}{C_n},$$

from which it follows that

$$1 + \frac{1}{n}(1 - qC_n) - q\frac{C_n}{n^2} \geqslant \frac{C_{n+1}}{C_n}.$$

Let us suppose first that, for some n,

$$1 - qC_n < 0.$$

In this case,

$$1 > 1 + \frac{1}{n}(1 - qC_n) - q\frac{C_n}{n^2} \geqslant \frac{C_{n+1}}{C_n},$$

so that $C_{n+1} < C_n$. If instead we have $1 - qC_n \geqslant 0$, that is, $C_n \leqslant \frac{1}{q}$, then, for all n,

$$C_n \leqslant \max\left\{C_1, \frac{1}{q}\right\}, \tag{1.19}$$

which proves the lemma.

Remark. We can show, although it would complicate the proof somewhat, that, if a sequence of positive numbers α_n satisfies the condition

$$\lambda_n - \lambda_{n+1} \geqslant q\lambda_n^p$$

where $1 < p < \infty$ and $q > 0$, then

$$\lambda_n = O\left(\frac{1}{n^{\frac{1}{p-1}}}\right).$$

Proof of Theorem 1.7: It follows from formula (1.16) that, for sufficiently large n,

$$0 \leqslant Fx_n(x_n - \bar{x}_n) \leqslant \nu\, (\delta_n)^{1/2},$$

where $\nu > 0$ and $\delta_n = f(x_n) - f(x_{n+1})$. If we set $\lambda_n = f(x_n) - f(y)$, where y is a point at which f attains a minimum on Ω, we see that

$$\delta_n = \lambda_n - \lambda_{n+1}.$$

Therefore, for sufficiently large n,

$$0 \leqslant Fx_n\left(x_n - \bar{x}_n\right) \leqslant \nu\,(\lambda_n - \lambda_{n+1})^{1/2}.$$

On the other hand, it follows from (1.9) that

$$\lambda_n \leqslant Fx_n\left(x_n - \bar{x}_n\right).$$

From the preceding inequalities, it follows that

$$\lambda_n - \lambda_{n+1} \geqslant \frac{1}{\nu^2}\lambda_n^2.$$

The assertion in the theorem now follows from Lemma 1.4.

Remark. It follows from (1.17) and (1.18) that, for sufficiently large n,

$$f(x_n)-f(y)=\lambda_n=\frac{C_n}{n}\leqslant\frac{1}{n}\max\left(f(x_1)-f(y),\ \frac{2}{\varepsilon_1\varepsilon_2}D\right).$$

Let us show that, in certain cases, the method we are considering converges with the speed of a geometric progression.

THEOREM 1.8. *Let X denote a Hilbert space and let Ω denote a convex bounded subset of X that satisfies the following condition: there exists a $\gamma>0$ such that, for any two points x and y in Ω, the point $\frac{x+y}{2}+z$, where $\|z\|\leqslant\gamma\|x-y\|^2$ also belongs to Ω. Suppose also, that a differentiable convex functional f has the property that*

$$\inf_{x\ni\Omega}\|Fx\|=\varepsilon>0.$$

Then, the sequence (1.12) converges with the speed of a geometric progression to the unique point representing a minimum of f on Ω.

Proof: It follows from the condition of the theorem that Ω is strictly convex and weakly compact. Therefore, f attains a minimum on Ω at a unique point y. Since the space X is a Hilbert space, the operator F maps X into X. Thus, Fx can be regarded either as an element of X or as a functional defined on X. We note[1] that

$$\|Fx\|^2=(Fx,\ Fx)=Fx\,(Fx).$$

Let us show first of all that there exists a number $q\in(0,\ 1)$ such that, for $n=1,\ 2,\ \ldots,$

$$\lambda_{n+1}\leqslant q\lambda_n, \tag{1.20}$$

where, just as before, $\lambda_n=f(x_n)-f(y)$. Let us consider two cases:

1) $\alpha_n=1$. By using (1.14) and (1.9), we obtain

$$\lambda_n-\lambda_{n+1}=\delta_n\geqslant\frac{\varepsilon_2}{2}Fx_n\left(x_n-\bar{x}_n\right)\geqslant\lambda_n\frac{\varepsilon_2}{2},$$

[1]By $(x,\ y)$, we denote the scalar product of elements $x,\ y\in X$.

so that

$$\lambda_{n+1} \leqslant \left(1 - \frac{\varepsilon_2}{2}\right)\lambda_n. \tag{1.21}$$

2) $\alpha_n < 1$. We note that, by virtue of our assumption regarding Ω, the element

$$\frac{x_n + \bar{x}_n}{2} - \gamma \frac{Fx_n}{\|Fx_n\|} \|\bar{x}_n - x_n\|^2 \in \Omega$$

is a member of Ω. By using this fact, we see that

$$\begin{aligned} Fx_n(\bar{x}_n - x_n) &= 2Fx_n\left(\frac{x_n + \bar{x}_n}{2} - \gamma \frac{Fx_n}{\|Fx_n\|} \|\bar{x}_n - x_n\|^2\right) + \\ &+ 2Fx_n\left(\gamma \frac{Fx_n}{\|Fx_n\|} \|\bar{x}_n - x_n\|^2 - x_n\right) \geqslant \\ &\geqslant 2Fx_n(\bar{x}_n) + 2\gamma \|Fx_n\| \cdot \|\bar{x}_n - x_n\|^2 - 2Fx_n(x_n) \geqslant \\ &\geqslant 2Fx_n(\bar{x}_n - x_n) + 2\gamma\varepsilon \|\bar{x}_n - x_n\|^2, \end{aligned}$$

so that

$$Fx_n(x_n - \bar{x}_n) \geqslant 2\gamma\varepsilon \|\bar{x}_n - x_n\|^2.$$

Now, using (1.15) and (1.19), we obtain

$$\begin{aligned} \lambda_n - \lambda_{n+1} = \delta_n &\geqslant \frac{\varepsilon_1\varepsilon_2}{2} \frac{[Fx_n(x_n - \bar{x}_n)]^2}{\|x_n - \bar{x}_n\|^2} = \\ &= \frac{\varepsilon_1\varepsilon_2}{2} \frac{Fx_n(x_n - \bar{x}_n)}{\|x_n - \bar{x}_n\|^2} Fx_n(x_n - \bar{x}_n) \geqslant \gamma\varepsilon\varepsilon_1\varepsilon_2\lambda_n, \end{aligned}$$

that is,

$$\lambda_{n+1} \leqslant (1 - \gamma\varepsilon\varepsilon_1\varepsilon_2)\lambda_n. \tag{1.22}$$

It follows from (1.21) and (1.22) that we may take for the q appearing in formula (1.20) the quantity

$$\max\left(1 - \frac{\varepsilon_2}{2},\ 1 - \gamma\varepsilon\varepsilon_1\varepsilon_2\right).$$

It follows from (1.20) that $\lambda_n \leqslant q^n \lambda_0$.

Let us use Theorem 4.3 of Chapter 1 to find a bound for $\|x_n - y\|$. In the present case, $\delta(\tau) = \gamma\tau^2$, where δ is the function mentioned in that theorem. By using formula (I.4.6) and the remark following that theorem, we obtain

$$\lambda_n \geqslant 2\|Fy\| \cdot \gamma \cdot \|x_n - y\|^2.$$

Thus,

$$\|x_n - y\| \leqslant \sqrt{\frac{\lambda_n}{2\|Fy\|\gamma}} = \sqrt{\frac{\lambda_0}{2\|Fy\|\gamma}} (q^{1/2})^n.$$

This completes the proof of the theorem.

6°. Consider the problem of finding the minimum on a convex compact set Ω of a functional f that is differentiable with respect to directions. A natural generalization of the conditional-gradient method to this case is the following algorithm: we choose and fix a number $a \in (0, 1]$.

1) We take as our first approximation an arbitrary element x_1 of Ω.

2) Suppose that the element x_n (for $n = 1, 2, \ldots$) is chosen. We find an element $\bar{x}_n$ of Ω such that

$$f'_{x_n}(\bar{x}_n) = \min_{x \in \Omega} f'_{x_n}(x),$$

and a number α_n with the property that

$$f(x_n + \alpha_n(\bar{x}_n - x_n)) = \min_{0 \leqslant \alpha \leqslant a_n} f(x_n + \alpha(\bar{x}_n - x_n)),$$

where a_n is an arbitrary number satisfying the inequalities

$$a \leqslant a_n \leqslant 1.$$

As our $(n+1)$st approximation, we take the element

$$x_{n+1} = x_n + \alpha_n(\bar{x}_n - x_n).$$

We shall say that a functional f is continuously differentiable with respect to directions on Ω if, for arbitrary $x \in \Omega$,

$$x_n \to x$$

and for arbitrary $u \in X$,

$$f'_{x_n}(u) \to f'_x(u).$$

THEOREM 1.9. *Suppose that Ω is a convex compact functional f that is continuously differentiable with respect to directions on Ω. Suppose that f'_x is, for every $x \in \Omega$, a continuous functional. Then, the necessary (and, in the case of convex f, sufficient) condition for a minimum (II.2.11) is satisfied at the cluster points of the sequence $\{x_n\}$ constructed above.*

The proof of Theorem 1.9 is carried out in the same way as the proof of Theorem 1.2.

7°. Unfortunately, many important functionals are not continuously differentiable. An example is the functional f considered in subsection 14°, Sec. 2, Chapter 2:

$$f(x) = \max_{t \in E} (Ax)(t), \tag{1.23}$$

where A is a completely continuous operator that is differentiable in the sense of Fréchet and that maps a normed space X into the space $C(E)$, where E is a closed bounded subset of R^n. We note that, under our assumptions regarding A, the functional (1.23) is continuously differentiable with respect to directions at every point x of X and that

$$f'_x(u) = \max_{t \in E(Ax)} (A'_x(u))(t),$$

where

$$E(Ax) = \{t \mid Ax(t) = \max_{\tau \in E} Ax(\tau)\}.$$

Let Ω denote a convex weakly compact subset of X. Then, we shall call $x \in \Omega$ a *stationary point* of the functional (1.23) if the necessary condition for a minimum (II.2.27)

$$\min_{y \in \Omega} \max_{t \in E(Ax)} (A'_x (y - x))(t) = 0$$

is satisfied at that point.

Let us give an algorithm that enables us in a certain sense to find stationary points of a functional f defined on Ω. This algorithm can be regarded as a generalization of the conditional-gradient method. Let ε' and ρ' denote arbitrary positive numbers.

1) We choose $x_1 \in \Omega$ arbitrarily.

2) Suppose that x_n (for $n = 1, 2, \ldots$) is already determined. Let us define

$$\delta_n = \min_{x \in \Omega} \max_{t \in E(Ax_n)} \left(A'_{x_n} (x - x_n) \right) (t).$$

If $\delta_n = 0$, the necessary condition (II.2.27) is satisfied and hence x_n is a stationary point. Let us now suppose that $\delta_n < 0$. Let us find the smallest natural number k and the element $\bar{x}_n$ for which

$$\min_{x \in \Omega} \max_{t \in E_{n, \frac{\varepsilon'}{2^k}}} \left(A'_{x_n} (x - x_n) \right) (t) =$$

$$= \max_{t \in E_{n, \frac{\varepsilon'}{2^k}}} \left(A'_{x_n} (\bar{x}_n - x_n) \right) (t) \leqslant -\rho'/2^k. \qquad (1.24)$$

Here

$$E_{n, \frac{\varepsilon'}{2^k}} = \left\{ t \in E \,\middle|\, |(Ax_n)(t) - \max_{\tau \in E} (Ax_n)(\tau)| < \frac{\varepsilon'}{2^k} \right\}.$$

Let us now define $x_{n+1} = x_n + \alpha_n (\bar{x}_n - x_n)$, where $\bar{x}_n$ is chosen in accordance with formula (1.24) and α_n is found from the condition

$$f(x_n + \alpha_n (\bar{x}_n - x_n)) = \min_{\alpha \in [0, 1]} f(x_n + \alpha (\bar{x}_n - x_n)). \qquad (1.25)$$

In what follows, we shall use the following notation: for an arbitrary natural number n, we define

$$\varepsilon_n = \frac{\varepsilon'}{2^k}, \quad \rho_n = \frac{\rho'}{2^k}, \qquad (1.26)$$

where k is the smallest natural number that, for given n, satisfies (1.24). For any $\varepsilon > 0$ and natural number n, we define

$$E_{n,\varepsilon} = \{t \in E \mid |(Ax_n)(t) - \max_{\tau \in E} (Ax_n)(\tau)| \leqslant \varepsilon\}. \tag{1.27}$$

THEOREM 1.10. *Let Ω denote a weakly compact set. Let A denote a completely continuous operator that is differentiable in the sense of Fréchet. Suppose that*

$$\sup_{x \in \Omega} \|A'_x\| < \infty.$$

Then, the sequence $\{x_n\}$ obtained by applying the alogrithm described above is such that, for arbitrary $\varepsilon > 0$,

$$\lim_{n \to \infty} \min_{x \in \Omega} \max_{t \in E_{n,\varepsilon}} \left(A'_{x_n}(x - x_n)\right)(t) = 0. \tag{1.28}$$

Proof: 1) Let us suppose first that

$$\lim_{n \to \infty} \rho_n = 0.$$

In this case, it follows from (1.26) that we also have

$$\lim_{n \to \infty} \varepsilon_n = 0.$$

Let us suppose the opposite. Then, again using (1.26), we can choose a subsequence $\{n_l\}_{l=1}^{\infty}$ with the following property: there exist positive numbers ε^* and ρ^* such that, for all l,

$$\varepsilon_{n_l} \geqslant 2\varepsilon^*, \quad \rho_{n_l} \geqslant 2\rho^*.$$

It follows from (1.24) that

$$\max_{t \in E_{n_l \varepsilon_{n_l}}} \left(A'_{x_{n_l}}\left(\bar{x}_{n_l} - x_{n_l}\right)\right)(t) \leqslant -2\rho^*. \tag{1.29}$$

Since the operator A is completely continuous, we can assume without loss of generality that there exists

$$\lim_l Ax_{n_l} = y.$$

Let us define

$$\widetilde{E}_{\varepsilon^*} = \{t \in E \mid |y(t) - \max_{\tau \in E} y(\tau)| < \varepsilon^*\}.$$

Since the sequence of the continuous function $(Ax_{n_l})(t)$ converges uniformly to a function $y(t)$ and $\varepsilon_{n_l} \geqslant 2\varepsilon^*$, it follows that, for sufficiently large l,

$$E_{n_l \varepsilon_{n_l}} \supset \widetilde{E}_{\varepsilon^*}.$$

It then follows from (1.29) that

$$\max_{t \in \widetilde{E}_{\varepsilon^*}} \left(A'_{x_{n_l}}\left(\bar{x}_{n_l} - x_{n_l}\right)\right)(t) \leqslant$$
$$\leqslant \max_{t \in E_{n_l}, \varepsilon_{n_l}} \left(A'_{x_{n_l}}\left(\bar{x}_{n_l} - x_{n_l}\right)\right)(t) \leqslant -2\rho^*. \tag{1.30}$$

From the formula on finite increments,

$$A\left(x_{n_l} + \alpha\left(\bar{x}_{n_l} - x_{n_l}\right)\right) = Ax_{n_l} + \alpha A'_{x_{n_l}}\left(\bar{x}_{n_l} - x_{n_l}\right) + o_{n_l}(\alpha), \tag{1.31}$$

where

$$\frac{\|o_{n_l}(\alpha)\|}{\alpha} \underset{\alpha \to 0}{\to} 0.$$

It follows from (1.30) and (1.31) that, for $\alpha > 0$,

$$\max_{t \in \widetilde{E}_{\varepsilon^*}} \left(A\left(x_{n_l} + \alpha\left(\bar{x}_{n_l} - x_{n_l}\right)\right)\right)(t) \leqslant \max_{t \in \widetilde{E}_{\varepsilon^*}} \left(Ax_{n_l}\right)(t) +$$
$$+ \alpha \max_{t \in \widetilde{E}_{\varepsilon^*}} \left(A'_{x_{n_l}}\left(\bar{x}_{n_l} - x_{n_l}\right)\right)(t) + \|o_{n_l}(\alpha)\| \leqslant$$
$$\leqslant \max_{t \in E} (Ax_{n_l})(t) - 2\alpha\rho^* + \|o_{n_l}(\alpha)\| =$$
$$= f(x_{n_l}) + \alpha\left(-2\rho^* + \frac{\|o_{n_l}(\alpha)\|}{\alpha}\right).$$

Remembering that the operator A is differentiable in the sense of Fréchet, there exists an $\alpha_1^* \in (0\ 1]$ such that

$$\frac{\|o_{n_l}(\alpha)\|}{\alpha} < \rho^*$$

for all $0 < \alpha < \alpha_1^*$ and sufficiently large l. For these α and l,

$$\max_{t \in \tilde{E}_{\varepsilon^*}} \left(A\left(x_{n_l} + \alpha\left(\bar{x}_{n_l} - x_{n_l}\right)\right)\right)(t) \leqslant f\left(x_{n_l}\right) - \alpha\rho^*. \tag{1.32}$$

It follows from the definition of the set $\tilde{E}_{\varepsilon^*}$ that, for sufficiently large l and for $t \in E \setminus \tilde{E}_{\varepsilon^*}$,

$$(Ax_{n_l})(t) \leqslant \max_{t \in E}(Ax_{n_l})(t) - \frac{\varepsilon^*}{2} = f(x_{n_l}) - \frac{\varepsilon^*}{2}.$$

Therefore, for these l and $\alpha > 0$

$$\sup_{t \in E \setminus \tilde{E}_{\varepsilon^*}} \left(A\left(x_{n_l} + \alpha\left(\bar{x}_{n_l} - x_{n_l}\right)\right)\right)(t) \leqslant \sup_{t \in E \setminus \tilde{E}_{\varepsilon^*}} (Ax_{n_l})(t) +$$
$$+ \alpha \sup_{t \in E \setminus \tilde{E}_{\varepsilon^*}} \left(A'_{x_{n_l}}\left(\bar{x}_{n_l} - x_{n_l}\right)\right)(t) + \|o_{n_l}(\alpha)\| \leqslant$$
$$\leqslant f(x_{n_l}) - \frac{\varepsilon^*}{2} + \alpha\left\|A'_{x_{n_l}}\left(\bar{x}_{n_l} - x_{n_l}\right)\right\| + \|o_{n_l}(\alpha)\|.$$

From the condition of the theorem,

$$\sup_l \left\|A'_{x_{n_l}}\left(\bar{x}_{n_l} - x_{n_l}\right)\right\| < \infty.$$

Remembering that A is differentiable in the sense of Fréchet, there exists an $\alpha_2^* > 0$ such that, for $0 < \alpha < \alpha_2^*$,

$$\alpha\left(\left\|A'_{x_{n_l}}\left(\bar{x}_{n_l} - x_{n_l}\right)\right\| + \frac{\|o_{n_l}(\alpha)\|}{\alpha}\right) < \frac{\varepsilon^*}{4}.$$

Thus, for sufficiently large l and $\alpha \in (0, \alpha_2^*]$,

$$\sup_{t \in E \setminus \tilde{E}_{\varepsilon^*}} \left(A\left(x_{n_l} + \alpha\left(\bar{x}_{n_l} - x_{n_l}\right)\right)\right)(t) \leqslant f(x_{n_l}) - \frac{\varepsilon^*}{4}. \tag{1.33}$$

Let us now define

$$\alpha^* = \min(\alpha_1^*, \alpha_2^*),$$

$$\sigma^* = \min\left(\alpha^*\rho^*, \frac{\varepsilon^*}{4}\right).$$

It follows from (1.32) and (1.33) that, for sufficiently large l and $\alpha < \alpha^*$,

$$f\left(x_{n_l} + \alpha\left(\bar{x}_{n_l} - x_{n_l}\right)\right) = \max_{t \in E}\left(A\left(x_{n_l} + \alpha\left(\bar{x}_{n_l} - x_{n_l}\right)\right)\right)(t) \leqslant \\ \leqslant f(x_{n_l}) - \sigma^*,$$

so that

$$f(x_{n_l+1}) = \min_{\alpha \in [0, 1]} f\left(x_{n_l} + \alpha\left(\bar{x}_{n_l} - x_{n_l}\right)\right) \leqslant f(x_{n_l}) - \sigma^*. \qquad (1.34)$$

It follows from the weak compactness of Ω and the complete continuity of A that f is bounded on Ω. Since (by construction) the sequence $\{f(x_n)\}$ decreases monotonically, there exists

$$\lim f(x_n) = C > -\infty.$$

Obviously, $f(x_n) = C + \eta_n$, where each $\eta_n > 0$ and $\eta_n \to 0$. Let us choose l so that $\eta_{n_l} < \frac{1}{2}\sigma^*$. Then, from (1.34),

$$f(x_{n_l+1}) \leqslant C + \eta_{n_l} - \sigma^* < C - \frac{1}{2}\sigma^*,$$

which contradicts the definition of C.

2) We have shown that

$$\varepsilon_n \to 0 \text{ and } \rho_n \to 0.$$

Keeping this in mind, let us proceed to prove the theorem. Since $x_n \in \Omega$ and $A'_{x_n}(0) = 0$, where $(n = 1, 2, \ldots)$, it follows that, for arbitrary positive ε and all n,

$$\min_{x \in \Omega} \max_{t \in E_{n,\varepsilon}} \left(A'_{x_n}(x - x_n)\right)(t) \leqslant 0. \tag{1.35}$$

Let us suppose that the theorem is not true. Then, on the basis of (1.35), there exist $\tilde{\varepsilon} > 0$, $\tilde{\rho} > 0$, and a subsequence $\{x_{n_i}\}$ such that

$$\min_{x \in \Omega} \max_{t \in E_{n_i, \tilde{\varepsilon}}} \left(A'_{x_{n_i}}\left(x - x_{n_i}\right)\right)(t) \leqslant -\tilde{\rho}.$$

Since $E_{n_i,\varepsilon} \subset E_{n_i,\tilde{\varepsilon}}$ for $\varepsilon < \tilde{\varepsilon}$, it follows that, for such ε,

$$\min_{x \in \Omega} \max_{t \in E_{n_i,\varepsilon}} \left(A'_{x_{n_i}}(x - x_{n_i})\right)(t) \leqslant -\tilde{\rho}. \tag{1.36}$$

However, this is impossible because, by definition $\rho_{n_i} = \frac{\rho'}{2^k}$, where $k = k(n_i)$ is the smallest natural number that satisfies (1.24). Since $\rho_{n_i} \to 0$, it follows that $k(n_i) \to \infty$. From this it follows, in particular, that, for sufficiently large i, we have $k(n_i) > 1$ and hence $k(n_i) - 1$ is a natural number. For this number, (1.24) is not valid; that is,

$$\min_{x \in \Omega} \max_{t \in E_{n_i, \frac{\varepsilon'}{2^{k(n_i)-1}}}} \left(A'_{x_{n_i}}\left(\bar{x}_{n_i} - x_{n_i}\right)\right)(t) > -\frac{\rho'}{2^{k(n_i)-1}}.$$

Since, for sufficiently large i,

$$\frac{\varepsilon}{2^{k(n_i)-1}} < \tilde{\varepsilon}, \quad \frac{\rho'}{2^{k(n_i)-1}} < \rho,$$

inequality (1.36) is not satisfied for these i. This completes the proof of the theorem.

2. The method of the projection of the gradient.

1°. Let H denote a Hilbert space and let Ω denote a convex closed bounded and hence weakly compact subset of H. Let x denote a member of H. Then, as we know, there exists a unique element $p_\Omega x$ of Ω such that

$$\|x - p_\Omega x\| = \min_{z \in \Omega} \|z - x\|.$$

The operator p defined above is called the ***projection operator*** onto Ω and the element $p_\Omega x$ is called the ***projection*** of the element x onto Ω. It follows from the definition of a projection that, for all $z \in \Omega$,

$$\|x - p_\Omega x\|^2 \leqslant \|z - x\|^2.$$

Since the functional $f^2(z) = \|z - x\|^2$ is strictly convex and differentiable and since its derivative $(f^2)'_x = 2(z - x)$, if we apply the necessary and sufficient condition for a minimum (II.2.12), we obtain the following result: the element $p_\Omega x$ is the projection of x onto Ω if and only if, for arbitrary $z \in \Omega$,

$$(p_\Omega x - x,\ p_\Omega x) \leqslant (p_\Omega x - x,\ z),$$

or, what amounts to the same thing, for arbitrary $z \in \Omega$,

$$(x - p_\Omega x,\ z - p_\Omega x) \leqslant 0. \tag{2.1}$$

Here, we shall consider a method of minimizing a differentiable function f defined on Ω, one that determines the direction of descent with the aid of the projection of the gradient. (In inequality (2.1) and the preceding inequality, the notation (x, y) denotes the scalar product of two members x and y of H. If g is a linear functional defined on H, then, by the familiar theorem of Riesz, $g \in H$ and, for arbitrary $x \in H$, we have $g(x) = (g, x)$.)

We point out that, in general, the problem of finding the projection of the gradient onto a convex closed bounded set is more complicated than minimizing the gradient on that set. Therefore, the direction of descent in the method considered below is found with greater difficulty than in the conditional-gradient method. However, we may assume that in solving a number of practical problems, this method yields better convergence.

2°. Let x denote a member of Ω and let g denote a member of H. Consider the ray

$$L = \{x_\alpha \mid x_\alpha = x - \alpha g,\ \alpha \in [0, \infty)\}.$$

Define

$$w_\alpha = p_\Omega x_\alpha. \tag{2.2}$$

If in (2.1) we set $x = x_\alpha$ and $p_\Omega x = w_\alpha$, we have, for arbitrary $z \in \Omega$

$$(x_\alpha - w_\alpha, z - w_\alpha) = (x - \alpha g - w_\alpha, z - w_\alpha) =$$
$$= -\alpha (g, z - w_\alpha) + x(z - w_\alpha) - (w_\alpha, z) + \| w_\alpha \|^2 \leqslant 0,$$

so that

$$2\alpha (g, w_\alpha - z) \leqslant -2 \| w_\alpha \|^2 + 2(w_\alpha, z) + 2(x, w_\alpha - z) =$$
$$= -(\| z \|^2 + \| w_\alpha \|^2 - 2(w_\alpha, z)) + \| z \|^2 - \| w_\alpha \|^2 + 2(x, w_\alpha - z).$$

Remembering that

$$\| z \|^2 + \| w_\alpha \|^2 - 2(w_\alpha, z) = \| z - w_\alpha \|^2,$$

we obtain finally

$$2\alpha (g, w_\alpha - z) \leqslant \| z \|^2 - \| w_\alpha \|^2 - \| w_\alpha - z \|^2 + 2(x, w_\alpha - z). \tag{2.3}$$

Formula (2.2) defines an abstract function w_α defined for $\alpha \in [0, \infty)$ with range in H.

THEOREM 2.1. *An abstract function w_α defined in accordance with formula (2.2) is continuous at an arbitrary point α' in $[0, \infty)$. The function $\psi(\alpha) = (g, w_\alpha)$ defined for $\alpha \geqslant 0$ is continuous and non-decreasing in its domain of definition.*

Proof: Let α' and α'' denote positive numbers. In (2.3), we first set $\alpha = \alpha'$ and $z = w_{\alpha''}$ and then set $\alpha = \alpha''$ and $z = w_{\alpha'}$, we obtain

$$2\alpha' (g, w_{\alpha'} - w_{\alpha''}) \leqslant \| w_{\alpha''} \|^2 - \| w_{\alpha'} \|^2 - \| w_{\alpha''} - w_{\alpha'} \|^2 +$$
$$+ 2(x, w_{\alpha'} - w_{\alpha''}),$$
$$2\alpha'' (g, w_{\alpha''} - w_{\alpha'}) \leqslant \| w_{\alpha'} \|^2 - \| w_{\alpha''} \|^2 - \| w_{\alpha'} - w_{\alpha''} \|^2 +$$
$$+ 2(x, w_{\alpha''} - w_{\alpha'}).$$

By adding these two inequalities, we obtain

$$\| w_{\alpha'} - w_{\alpha''} \| \leqslant (\alpha' - \alpha'')(g, w_{\alpha''} - w_{\alpha'}).$$

Since Ω is bounded and $w_\alpha \in \Omega$ for $\alpha \geqslant 0$, it follows that the quantity

$$\sup_{\alpha'>0,\, \alpha''>0} |(g,\ w_{\alpha''} - w_{\alpha'})|$$

is finite. Therefore, it follows from the inequality obtained above that

$$\lim_{\alpha'' \to \alpha'} w_{\alpha''} = w_{\alpha'}.$$

From the same inequality, it follows that, for $\alpha' > \alpha''$,

$$(g,\ w_{\alpha'}) \leqslant (g,\ w_{\alpha''}).$$

Thus, the function $\psi(\alpha) = (g,\ w_\alpha)$ is nonincreasing for $\alpha \geqslant 0$. We note that the function $\psi(\alpha)$ will decrease on the interval $[\alpha_1,\ \alpha_2]$ if $w_{\alpha'} \neq w_{\alpha''}$ for $\alpha',\ \alpha'' \in [\alpha_1,\ \alpha_2]$. The continuity of $\psi(\alpha)$ follows from the continuity of the function w_α.

THEOREM 2.2. *Suppose that the abstract function w_α is defined in accordance with formula (2.2). Then, there exists*

$$u = \lim_{\alpha \to \infty} w_\alpha,$$

and[1] $u \in G_\Omega g$.

Proof: Suppose that $\alpha_k \to \infty$. Without loss of generality, we may assume that there exists an element u in Ω such that $w_{\alpha_k} \overset{w}{\to} u$. If we set $\alpha = \alpha_k$ in (2.3), we obtain, for arbitrary $z \in \Omega$,

$$2\alpha_k\left(g,\ w_{\alpha_k} - z\right) \leqslant \|z\|^2 - \|w_{\alpha_k}\|^2 - \|w_{\alpha_k} - z\|^2 + \qquad (2.4)$$
$$+ 2\left(x,\ w_{\alpha_k} - z\right).$$

Since Ω is bounded, the right-hand member of this inequality fails, for all α_k, to exceed some particular number. Since $\alpha_k \to \infty$, inequality (2.4) holds only when $\lim\ (g,\ w_{\alpha_k} - z) = (g,\ u - z) \leqslant 0$. It then follows that $u \in G_\Omega g$. Let us show that $w_{\alpha_k} \to u$. Let us suppose that this is not the case. We may assume that $\|w_{\alpha_k} - u\| \geqslant \rho > 0$ for infinitely many values of k. Since $w_{\alpha_k} \overset{w}{\to} u$, the inequalities

[1]The set $G_\Omega g$ is defined in accordance with formula (I.5.1): an element z belongs to $G_\Omega g$ if and only if $z \in \Omega$ and $(g,\ z) = \min_{y \in \Omega} (g,\ y)$.

$$\|u\| < \|w_{\alpha_k}\| + \frac{\rho}{4}, \qquad (x, w_{\alpha_k} - u) < \frac{\rho}{8}$$

are satisfied for sufficiently large k. By setting $z = u$ in (2.4), we have, for these k,

$$2\alpha_k (g, w_{\alpha_k} - u) \leqslant \|u\|^2 - \|w_{\alpha_k}\|^2 - \|w_{\alpha_k} - u\|^2 + 2(x, w_{\alpha_k} - u) \leqslant \leqslant \frac{\rho}{4} - \rho + \frac{\rho}{4} = -\frac{\rho}{2}.$$

This last relation contradicts the fact that $u \in G_\varrho g$ and shows that $u = \lim w_{\alpha_k}$.

To complete the proof of the theorem, it remains to show that the element u is independent of the choice of sequence $\{\alpha_k\}$. Again let us suppose the opposite. Let $\{\alpha_k\}$ and $\{\beta_i\}$ denote two sequences that approach ∞ and suppose that

$$\lim w_{\alpha_k} = u' \neq u'' = \lim w_{\beta_i}.$$

Since $u' \in G_\varrho g$ and $u'' \in G_\varrho g$, we have $(x - u', g) = (x - u'', g) \geqslant 0$. One can easily show that, for

$$\alpha_0 = \frac{(x - u', g)}{\|g\|^2} = \frac{(x - u'', g)}{\|g\|^2} \geqslant 0$$

the elements $x_{\alpha_0} - u'$ and $x_{\alpha_0} - u''$ are orthogonal to the element g. Two cases are possible:

1) $\|u' - x_{\alpha_0}\| = \|u'' - x_{\alpha_0}\|$. Obviously, in this case, for arbitrary $\alpha \in [0, \infty)$,

$$\|x_\alpha - u'\|^2 = \|(x_\alpha - x_{\alpha_0}) + (x_{\alpha_0} - u')\|^2 = \|(\alpha - \alpha_0) g + (x_{\alpha_0} - u')\|^2 = = |\alpha - \alpha_0|^2 \|g\|^2 + \|x_{\alpha_0} - u'\|^2 = = |\alpha - \alpha_0|^2 \|g\|^2 + \|x_{\alpha_0} - u''\|^2 = \|x_\alpha - u''\|^2.$$

Using this fact, we can easily show by direct verification that, for all k,

$$\|u' - x_{\alpha_k}\|^2 = \frac{1}{2}(\|u' - x_{\alpha_k}\|^2 + \|u'' - x_{\alpha_k}\|^2) = = \frac{1}{4}\|u' - u''\|^2 + \left\|\frac{1}{2}(u' + u'') - x_{\alpha_k}\right\|^2.$$

Since $w_{\alpha_k} \to u'$ and, by virtue of our assumption, $\|u' - u''\|^2 > 0$, it follows that, for sufficiently large k,

$$\left\| \frac{1}{2}(u' + u'') - x_{\alpha_k} \right\| < \| w_{\alpha_k} - x_{\alpha_k} \|,$$

which contradicts the definition of w_{α_k}.

2) $\|u' - x_{\alpha_0}\| \neq \|u'' - x_{\alpha_0}\|$. Suppose, for example, that $\|u' - x_{\alpha_0}\| < < \|u'' - x_{\alpha_0}\|$. Then, remembering that the vectors $u'' - x_{\alpha_0}$ and $x_{\alpha_0} - x_{\beta_i}$ are orthogonal, we have

$$\|u'' - x_{\beta_i}\|^2 = \|u'' - x_{\alpha_0} + x_{\alpha_0} - x_{\beta_i}\|^2 = \|u'' - x_{\alpha_0}\|^2 + \|x_{\alpha_0} - x_{\beta_i}\|^2,$$

from which follows that

$$\|w_{\beta_i} - x_{\beta_i}\|^2 = \|u'' - x_{\alpha_0}\|^2 + \|x_{\alpha_0} - x_{\beta_i}\|^2 + \varepsilon_i,$$

where

$$\lim_{i \to \infty} \varepsilon_i = 0.$$

By using the relation

$$\|u' - x_{\beta_i}\|^2 = \|u' - x_{\alpha_0}\|^2 + \|x_{\alpha_0} - x_{\beta_i}\|^2,$$

which is valid by virtue of the orthogonality of the vectors $u' - x_{\alpha_0}$ and $x_{\alpha_0} - x_{\beta_i}$, we see that, in the present case,

$$\|u' - x_{\beta_i}\|^2 = \|u' - x_{\alpha_0}\|^2 + \|x_{\alpha_0} - x_{\beta_i}\|^2 < \|u'' - x_{\alpha_0}\|^2 + \\ + \|x_{\alpha_0} - x_{\beta_i}\|^2 = \|w_{\beta_i} - x_{\beta_i}\|^2 - \varepsilon_i,$$

from which it follows that, for sufficiently large i,

$$\|u' - x_{\beta_i}\|^2 < \|w_{\beta_i} - x_{\beta_i}\|^2.$$

This inequality contradicts the definition of w_{β_i} and completes the proof of the theorem.

THEOREM 2.3. *If* $\alpha > 0$, *then*

$$(g, w_\alpha - x) \leqslant -\frac{\|x - w_\alpha\|^2}{\alpha}. \tag{2.5}$$

Proof: By setting $z=x$ in (2.3), we obtain

$$2\alpha(g,\ w_\alpha - x) \leqslant \|x\|^2 - \|w_\alpha\|^2 - \|w_\alpha - x\|^2 + 2(x,\ w_\alpha - x) =$$
$$= -(\|x\|^2 + \|w_\alpha\|^2 - 2(x,\ w_\alpha)) - \|w_\alpha - x\|^2 = -2\|x - w_\alpha\|^2,$$

from which (2.5) follows. This completes the proof of the theorem.

THEOREM 2.4. *If* $(g,\ w_{\alpha_1} - x) = 0$ *for some* $\alpha_1 > 0$, *then either* $g=0$ *or* $x \in G_\Omega g$.

Proof: Let us suppose that $g \neq 0$. It follows from (2.5) that, in this case, $x = w_{\alpha_1}$. It now follows from (2.1) that, for all $z \in \Omega$,

$$(x_{\alpha_1} - x,\ z - x) \leqslant 0.$$

Then, since $x - x_{\alpha_1} = \alpha_1 g$, we have

$$(g,\ x) = \min_{z \in \Omega} (g,\ z),$$

as we wished to show.

Let us look at some corollaries of these theorems.

COROLLARY 2.1. *Suppose that* $(g,\ w_{\alpha_1} - x) < 0$ *for at least one* $\alpha_1 \in (0,\ \infty)$. *Then,* $(g,\ w_\alpha - x) < 0$ *for all* $\alpha \in (0,\ \infty)$.

Proof: It follows from (2.5) that $(g,\ w_\alpha - x) \leqslant 0$. If $(g,\ w_{\alpha_2} - x) = 0$ for some α_2, then, by virtue of Theorem 2.4, we have $x \in G_\Omega g$. Since (by virtue of Theorem 2.2) $u \in G_\Omega g$, it follows that $(g,\ u - x) = 0$. It follows from Theorem 2.1 that the function $\varphi(\alpha) = (g,\ w_\alpha - x)$ does not increase. Since $\varphi(0) = 0$ (obviously $w_0 = x$) and

$$\lim_{\alpha \to \infty} \varphi(\alpha) = (g,\ u - x) = 0,$$

it follows that $\varphi(\alpha) = 0$ for every $\alpha \geqslant 0$. This contradicts the fact that $\varphi(\alpha_1) < 0$.

COROLLARY 2.2. *If* $(g,\ u - x) < 0$, *then* $(g,\ w_\alpha - x) < 0$ *for all* $\alpha \in (0,\ \infty)$.

Proof: It follows from Theorem 2.2 that

$$(g,\ u - x) = \lim_{\alpha \to \infty} (g,\ w_\alpha - x).$$

This means that, for sufficiently large α_1,

$$(g,\ w_{\alpha_1} - x) < 0,$$

and, hence, for all α,

$$(g, w_\alpha - x) < 0.$$

COROLLARY 2.3. *If* $g \neq 0$ *and* $(g, w_{\alpha_1} - x) = 0$ *for some* $\alpha_1 > 0$, *then* $(g, u - x) = 0$ *and, for every* $\alpha > 0$, *we have* $x = w_\alpha$.

Proof: It follows easily from Corollary 2.1 that, in this case, $(g, w_\alpha - x) = 0$ for all $\alpha \geqslant 0$. It follows from this that

$$(g, u - x) = \lim_{\alpha \to 0} (g, w_\alpha - x) = 0.$$

By applying formula (2.5), we see that $x = w_\alpha$ for $\alpha > 0$.

In the same way, we can prove

COROLLARY 2.4. *If* $(g, u - x) = 0$, *then, for arbitrary* $\alpha > 0$, *we have* $x = w_\alpha$.

3°. Let f denote a differentiable functional defined on H and let x denote a member of Ω. Define $g = Fx$ (where F is the gradient of the functional f). Also define

$$u_x = \lim_{\alpha \to \infty} w_\alpha.$$

(In the preceding subsection, this element was denoted by u.) Since (by virtue of Theorem 2.2) $u_x \in G_\Omega Fx$, the necessary (and also sufficient in case f is convex) condition for a minimum (II.2.12) of the functional f on Ω can be written

$$(Fx, u_x - x) = 0. \tag{2.6}$$

Let us set $w_\alpha(x) = p_\Omega(x - \alpha Fx)$ [this element was denoted in the preceding subsection by w_α]. It follows from Corollary 2.1 - 2.4 that satisfaction of condition (2.6) is equivalent to satisfaction of the equation $w_\alpha(x) = x$ for some (and hence every) $\alpha > 0$. Conversely, nonsatisfaction of this condition is equivalent to the inequality

$$(Fx, w_\alpha(x) - x) < 0 \text{ for some } \alpha < 0.$$

In particular, the definition of a stationary point that was given in Sec. 1 can be reformulated in new terms as follows: a point y is a stationary point of f on Ω if $y = w_\alpha(y)$ for some $\alpha > 0$.

4°. The overall procedure of the method of projection of the gradient consists in the following: as our first approximation x_1, we choose an arbitrary element in Ω. Once the nth approximation x_n is chosen, if x_n is not a stationary point, we take

$$x_{n+1} = x_n + \beta_n (w_{\alpha_n}(x_n) - x_n),$$

where β_n is chosen in some definite method from the interval [0, 1] and α_n from the interval $[0, \infty]$. Let us write $w_\infty(x_n) = u_{x_n}$. When we decide on a definite procedure for choosing the parameters xx and xx, we have a specific scheme for the method of projection of the gradient. Several such schemes are exhibited in [15]. Let us look at four specific schemes of this method.

We shall assume that f is a differentiable functional whose gradient, the operator F, satisfies a Lipschitz condition (1.5) on the set Ω: for $x', x'' \in \Omega$,

$$\| Fx' - Fx'' \| < L \| x' - x'' \|.$$

The first scheme of the method.

Let α_0, α', b, and a denote numbers such that

$$0 \leqslant \alpha_0 < \alpha' < \infty, \; 0 < b \leqslant 1, \; 0 \leqslant a < \min\left(b, \frac{2}{L\alpha'}\right). \tag{2.7}$$

For $n = 1, 2, \ldots$, let us define

$$x_{n+1} = x_n + \beta_n (w_{\alpha_n}(x_n) - x_n),$$

where $\alpha_0 < \alpha_n \leqslant \alpha'$ and β_n is chosen from the condition

$$f(x_n + \beta_n (w_{\alpha_n}(x_n) - x_n)) = \min_{a_n \leqslant \beta \leqslant b_n} f(x_n + \beta (w_{\alpha_n}(x_n) - x_n))$$
$$(b \leqslant b_n \leqslant 1, \; 0 \leqslant a_n \leqslant a).$$

Remark. When $a = 0$, if we let the number α_0 increase without bound, this scheme of the method of projection of the gradient "approaches" the conditional-gradient method.

The second scheme of the method.

Let α_0, α', and β_0 denote numbers such that

$$0 \leqslant \alpha_0 < \alpha' < \frac{2}{L}, \; 0 < \beta_0 \leqslant 1. \tag{2.8}$$

For $n = 1, 2, \ldots$, let us set

$$x_{n+1} = x_n + \beta_n (w_{\alpha_n}(x_n) - x_n),$$

where $\alpha_0 < \alpha_n \leqslant \alpha'$ and $\beta_0 \leqslant \beta_n \leqslant 1$.

5°. Let us look at the question of convergence of these schemes of the method of projection of the gradient. This question has been studied in detail in [39] for the second scheme in the case when $\beta_n = 1$ and

$$x_{n+1} = w_{\alpha_n}(x_n)$$

for all n. We shall give the proofs using approximately the same line of reasoning as in [39].

THEOREM 2.5. *Let $\{x_n\}$ denote a sequence constructed either according to the first or the second scheme shown above. Then, the sequence $\{f(x_n)\}$ decreases monotonically and*

$$\lim_{n\to\infty} \| x_n - w_{\alpha_n}(x_n) \| = 0.$$

Proof: By using formulas (1.6) and (2.5), we obtain, for $\beta \geqslant 0$,

$$f(x_n + \beta(w_{\alpha_n}(x_n) - x_n)) \leqslant f(x_n) + \beta(Fx_n, w_{\alpha_n}(x_n) - x_n) + \\ + \frac{\beta^2}{2} L \| w_{\alpha_n}(x_n) - x_n \|^2 \leqslant f(x_n) - \frac{\beta}{\alpha_n} \| x_n - w_{\alpha_n}(x) \|^2 + \\ + \frac{\beta^2}{2} L \| x_n - w_{\alpha_n}(x_n) \|^2 = f(x_n) - \left(\frac{\beta}{\alpha_n} - \frac{\beta^2}{2} L \right) \| x_n - w_{\alpha_n}(x_n) \|^2. \quad (2.9)$$

Let us suppose first that $\{x_n\}$ is the sequence constructed according to the first scheme. Then, for $\beta \in [a, b]$, we obtain by using (2.9)

$$f(x_n) - f(x_{n+1}) \geqslant f(x_n) - f(x_n + \beta(w_{\alpha_n}(x_n) - x_n)) \geqslant \\ \geqslant \left(\frac{\beta}{\alpha_n} - \frac{\beta^2}{2} L \right) \| x_n - w_{\alpha_n}(x_n) \|^2 \geqslant \left(\frac{\beta}{\alpha'} - \frac{\beta^2}{2} L \right) \| x_n - w_{\alpha_n}(x_n) \|^2.$$

Suppose that

$$a < \beta_0 < \min\left(\frac{2}{La'}, b \right).$$

Then,

$$\varepsilon_1 = \frac{\beta_0}{\alpha'} - \frac{\beta_0^2}{2} L > 0.$$

Thus, in the present case,

$$f(x_n)-f(x_{n+1})\geqslant \varepsilon_1\|x_n-w_{\alpha_n}(x_n)\|^2. \quad (2.10)$$

Now suppose that the sequence $\{x_n\}$ is constructed according to the second scheme. If we set $\beta=\beta_n$ in (2.9), we have

$$f(x_n)-f(x_{n+1})\geqslant \beta_n\left(\frac{1}{\alpha_n}-\frac{\beta_n}{2}L\right)\|x_n-w_{\alpha_n}(x_n)\|^2\geqslant$$
$$\geqslant \beta_0\left(\frac{1}{\alpha'}-\frac{L}{2}\right)\|x_n-w_{\alpha_n}(x_n)\|^2.$$

In view of the boundedness imposed on α',

$$\varepsilon_2=\beta_0\left(\frac{1}{\alpha'}-\frac{L}{2}\right)=\beta_0\frac{2-L\alpha'}{2\alpha'}>0.$$

Thus, in this case,

$$f(x_n)-f(x_{n+1})\geqslant \varepsilon_2\|x_n-w_{\alpha_n}(x_n)\|^2. \quad (2.11)$$

It follows from (2.10) and (2.11) that the sequence $\{f(x_n)\}$ decreases monotonically whether constructed according to the first or the second scheme. Since, on the basis of Lemma 1.1, the functional f is bounded on Ω, this sequence has a limit. That is,

$$\lim\|x_n-w_{\alpha_n}(x_n)\|=0$$

now follows immediately from (2.10) and (2.11). This completes the proof of the theorem.

Remark 1. Since

$$x_{n+1}-x_n=\beta_n(w_{\alpha_n}(x_n)-x_n),$$

it follows from the theorem that $\|x_{n+1}-x_n\|\to 0$.

Remark 2. One can combine the two schemes by taking one approximation according to the first scheme and the next according to the other. When one does this, it is easy to see from the proofs given above that Theorem 2.5 remains valid for the sequence so constructed.

THEOREM 2.6. *Let f denote a convex functional that attains a minimum on Ω at the point y. Let $\{x_n\}$ denote the sequence constructed according to the first (resp. second) scheme and suppose that the number α_0 appearing in (2.7) [resp. (2.8)] is positive. Then,*

$$f(x_n)-f(y)=O\left(\frac{1}{n}\right).$$

Proof: By applying Lagrange's formula and using the convexity of f, we have [see (II.3.22)]

$$f(y)-f(x_n)=(F(x_n+\theta(y-x_n)),\ y-x_n)\geqslant(Fx_n,\ y-x_n).$$

Let us define $\lambda_n=f(x_n)-f(y)$. From (2.1), we obtain

$$\begin{aligned}\lambda_n&\leqslant(Fx_n,\ x_n-y)=(Fx_n,\ x_n-w_{\alpha_n}(x_n))+(Fx_n,\ w_{\alpha_n}(x_n)-y)=\\&=(Fx_n,\ x_n-w_{\alpha_n}(x_n))+\frac{1}{\alpha_n}(x_n-\alpha_nFx_n-w_{\alpha_n}(x_n),\ y-w_{\alpha_n}(x_n))-\\&-\frac{1}{\alpha_n}(x_n-w_{\alpha_n}(x_n),\ y-w_{\alpha_n}(x_n))\leqslant\|Fx_n\|\cdot\|x_n-w_{\alpha_n}(x_n)\|+\\&+\frac{1}{\alpha_n}\|x_n-w_{\alpha_n}(x_n)\|\cdot\|y-w_{\alpha_n}(x_n)\|\leqslant\\&\leqslant\left(\|Fx_n\|+\frac{1}{\alpha_0}\|y-w_{\alpha_n}(x_n)\|\right)\|x_n-w_{\alpha_n}(x_n)\|.\end{aligned}$$

On the basis of Lemma 1.1, we have

$$k=\sup_{x\in\Omega}\|Fx\|<\infty.$$

Thus,

$$\lambda_n\leqslant\left(k+\frac{1}{\alpha_0}D\right)\|x_n-w_{\alpha_n}(x_n)\|,$$

where D is the diameter of Ω. On the other hand, it follows from (2.10) and (2.11) that

$$\lambda_n-\lambda_{n+1}=f(x_n)-f(x_{n+1})\geqslant\varepsilon_i\|x_n-w_{\alpha_n}(x_n)\|^2$$

where $i=1$ or 2 according as the first or second scheme is chosen. Combining the last two inequalities, we have

$$\lambda_n-\lambda_{n+1}\geqslant\frac{\varepsilon_i}{\left(k+\frac{1}{\alpha_0}D\right)^2}\lambda_n^2.$$

To complete the proof, it only remains to use Lemma 1.4.

Remark 1. It follows from (1.19), that, when we use the first scheme,

$$\lambda_n = f(x_n) - f(y) \leqslant \frac{1}{n} \max\left(f(x_1) - f(y), \frac{\left(k + \frac{1}{\alpha_0} D\right)^2}{\varepsilon_1(\beta)}\right),$$

where

$$\varepsilon_1(\beta) = \frac{\beta}{\alpha'} - \frac{\beta^2}{2} L,$$

and β satisfies only the inequalities

$$a < \beta < \min\left(\frac{2}{L\alpha'}, b\right).$$

Let us suppose that $a \leqslant \frac{1}{L\alpha'} \leqslant b$. For $\beta = \frac{1}{L\alpha'}$, the function $\varepsilon_1(\beta)$ assumes the largest value in its domain of definition, which is equal to $\frac{1}{2L\alpha'}$. Thus, if our assumption is satisfied, we have

$$f(x_n) - f(y) \leqslant \frac{1}{n} \max\left(f(x_1) - f(y), 2L\alpha'\left(k + \frac{1}{\alpha_0} D\right)^2\right).$$

When one uses the second scheme and remembers that

$$\varepsilon_2 = \beta_0 \frac{2 - L\alpha'}{2\alpha'},$$

we have

$$f(x_n) - f(y) \leqslant \frac{1}{n} \max\left(f(x_1) - f(y), \frac{2\alpha'\left(k + \frac{1}{\alpha_0} D\right)^2}{\beta_0(2 - L\alpha')}\right). \tag{2.12}$$

Remark 2. Under the assumptions of Theorem 2.6, the sequence $\{x_n\}$ is a minimizing sequence. Therefore, in the present case, the theorems of subsection 2° of Sec. 4 of Chapter 1 are applicable. In particular, the maximum that we are seeking is attained at the cluster points (in the weak sense) of the sequence in question. This sequence converges weakly if f is a strongly convex functional and it converges in norm if the conditions of Theorem 4.2 or 4.3 of Chapter 1 are satisfied.

Remark 3: It is shown in [39] that, if f is a twice differentiable functional and if $m\|u\|^2 \leqslant (F'x(u), u) \leqslant L\|u\|^2$ (where $m > 0$) for all x in Ω and all u in H, then the sequence $\{x_n\}$ constructed according to the second scheme with $\alpha_n = \alpha$, where $0 < \alpha < \frac{2}{L}$ and $\beta_n = 1$ for $n = 1, 2,$ converges to a point minimizing f on Ω with the speed of a geometric progression.

6°. Let us now look at the third scheme of projection of the gradient.

Suppose that

$$\varepsilon > 0 \text{ and } 0 < \alpha' \leqslant \frac{2}{2\varepsilon + L}. \qquad (2.13)$$

Let us set

$$x_{n+1} = w_{\alpha_n}(x_n),$$

where α_n is chosen from the condition

$$f(w_{\alpha_n}(x_n)) = \min_{a_n \leqslant \alpha \leqslant b_n} f(w_\alpha(x_n))$$
$$(0 \leqslant a_n \leqslant \alpha', \ \alpha' \leqslant b_n \leqslant \infty).$$

THEOREM 2.7. *Let $\{x_n\}$ denote a sequence constructed according to the third method. Then, the sequence $\{f(x_n)\}$ decreases monotonically and*

$$\lim_{n\to\infty} \|x_n - w_{\alpha'}(x_n)\| = 0.$$

Proof: By using formulas (1.6) and (2.5), we obtain, for $\alpha' \in [a_n, b_n]$,

$$f(w_\alpha(x_n)) = f(x_n + (w_\alpha(x_n) - x_n)) \leqslant f(x_n) + (Fx_n, w_\alpha(x_n) - x_n) +$$
$$+ \frac{L}{2}\|w_{\alpha_n}(x_n) - x_n\|^2 \leqslant f(x_n) - \frac{1}{\alpha}\|x_n - w_\alpha(x_n)\|^2 +$$
$$+ \frac{L}{2}\|x_n - w_\alpha(x_n)\|^2.$$

By using the definition of α_n and the fact that $\alpha \in [a_n, b_n]$, we obtain

$$f(w_{\alpha_n}(x_n)) = f(x_{n+1}) \leqslant f(x_n) + \left(\frac{L}{2} - \frac{1}{\alpha'}\right)\|x_n - w_{\alpha'}(x_n)\|^2,$$

from which we obtain by using (2.13)

$$f(x_n)-f(x_{n+1})\geqslant\left(\frac{1}{\alpha'}-\frac{L}{2}\right)\|x_n-w_{\alpha'}(x_n)\|^2\geqslant\varepsilon\|x_n-w_{\alpha'}(x_n)\|^2. \quad (2.14)$$

It follows from (2.14) that the sequence $\{f(x_n)\}$ decreases monotonically and hence has a limit. This means that

$$\lim_n\|x_n-w_{\alpha'}(x_n)\|^2=0.$$

This completes the proof of the theorem.

Reasoning as in the proof of Theorem 2.6, we can easily show that convexity of f implies that

$$f(x_n)-Q=O\left(\frac{1}{n}\right)$$

where

$$Q=\min_{x\in\Omega}f(x)\Big).$$

7°.

The fourth scheme

Let α', ε_1, and ε_2 denote given numbers such that

$$\alpha'>0,\quad 0<\varepsilon_1\leqslant\frac{2}{L},\quad 0<\varepsilon_2\leqslant 2-L\varepsilon_1.$$

We define

$$x_{n+1}=x_n+\beta_n(w_{\alpha_n}(x_n)-x_n),$$

where

$$0<\alpha_n\leqslant\alpha',\ \beta_n=\min\left(1,\ \gamma_n\frac{(Fx_n,\ x_n-w_{\alpha_n}(x_n))}{\|x_n-w_{\alpha_n}(x_n)\|^2}\right),\ \varepsilon_1\leqslant\gamma_n\leqslant\frac{2-\varepsilon_2}{L}.$$

THEOREM 2.8. *Let* $\{x_n\}$ *denote the sequence obtained according to the fourth scheme. Then the sequence* $\{f(x_n)\}$ *decreases monotonically and*

$$\lim\|x_n-w_{\alpha_n}(x_n)\|=0.$$

Proof: Reasoning as in the proof of Theorem 1.6, we can easily show that

$$\lim_{n\to\infty}(Fx_n,\ x_n - w_{\alpha_n}(x_n)) = 0.$$

Since, by virtue of (2.5),

$$(Fx_n,\ x_n - w_{\alpha_n}(x_n)) \geqslant \frac{\|x_n - w_{\alpha_n}(x_n)\|^2}{\alpha_n} \geqslant \frac{1}{\alpha'}\|x_n - w_{\alpha_n}(x_n)\|^2,$$

we have

$$\|x_n - w_{\alpha_n}(x_n)\|^2 \to 0.$$

That the sequence $\{f(x_n)\}$ decreases monotonically can be shown in the same way as in Theorem 1.6. This completes the proof of the theorem.

If f is a convex functional, then, as one can easily show,

$$f(x_n) - Q = O\left(\frac{1}{n}\right)$$

with this scheme just as with the preceding one.

8°. We conclude this section with a method similar to the method of projection of the gradient that is useful for minimizing a differentiable function f on a weakly compact but not necessarily convex subset Ω of H. We note that, in this case, for every $x \in H$ there exists an element (not necessarily unique) $p_\Omega x$ of Ω such that

$$\|p_\Omega x - x\| = \min_{y\in\Omega}\|y - x\|.$$

We shall assume that the functional f is differentiable in the sense of Fréchet and that its gradient F satisfies the Lipschitz condition (1.5) on the convex hull of the set Ω.

Let x denote a member of Ω and let α denote a nonnegative number. Let $p_\alpha(x)$ denote the set of all elements Ω lying at the minimum distance from $x - \alpha Fx$. In other words, the element $w_\alpha(x)$ belongs to $p_\alpha(x)$ if and only if $w_\alpha(x) \in \Omega$ and

$$\|w_\alpha(x) - (x - \alpha Fx)\| = \min_{z\in\Omega}\|z - (x - \alpha Fx)\|.$$

It follows from this equation that, for every z in Ω and every $w_\alpha(x)$ in $p_\alpha(x)$,

$$\|(w_\alpha(x)-x)+\alpha Fx\|^2 \leqslant \|(z-x)+\alpha Fx\|^2,$$

from which it follows that

$$\|w_\alpha(x)-x\|^2+2\alpha(Fx,\ w_\alpha(x)-x) \leqslant \|z-x\|^2+2\alpha(Fx,\ z-x).$$

Let us rewrite this last inequality in the form

$$2\alpha(Fx,\ w_\alpha(x)-z) \leqslant \|z-x\|^2-\|w_\alpha(x)-x\|^2. \tag{2.15}$$

In the present case, inequality of (2.15) is the analog of (2.3). In particular, for $z=x$ and $\alpha>0$, inequality (2.15) implies the inequality

$$(Fx,\ w_\alpha(x)-z) \leqslant -\frac{\|w_\alpha(x)-x\|^2}{2\alpha}, \tag{2.16}$$

which is the analog of (2.5).

We shall call a point $x \in \Omega$ a stationary point of the functional f defined on the set Ω if the necessary condition for a minimum (II.2.3) is satisfied at that point.

THEOREM 2.9. *Let x denote a point in Ω such that $x \in p_\alpha(x)$ for some $\alpha>0$. Then, x is a stationary point of f on Ω.*

Proof: Let u denote a member of $M_x(\Omega)$. Then, from the definition of the cone $M_x(\Omega)$, there exists a sequence of members u_s of H and a numerical sequence $\{\alpha_s\}$ such that

1) $u_s \to u$,
2) $\alpha_s > 0,\ \alpha_s \to 0$,
3) $x+\alpha_s u_s \in \Omega$.

If we set $w_\alpha(x)=x$ and $z=x+\alpha_s u_s$ in (2.15), we obtain

$$-2\alpha(Fx,\ \alpha_s u_s) \leqslant \alpha_s^2\|u_s\|^2.$$

Division by α_s yields the result

$$\alpha_s\|u_s\|^2+2\alpha(Fx,\ u_s) \geqslant 0.$$

Since $u_s \to u$, the sequence $\{u_s\}$ is bounded. Therefore, if we take the limit as $s \to \infty$, we see that $(Fx, u) \geqslant 0$. This completes the proof of the theorem.

To minimize the functional f on the set Ω, we can apply alogrithms analogous to the third scheme of the method of projection of the gradient and the second scheme of that method with $\beta_0 = 1$. We shall consider only the method analogous to the second scheme.

Let α' denote a number in the interval $\left(0, \frac{1}{L}\right)$, where L is defined by formula (1.5). As our first approximation, let us choose an arbitrary element x_1 of Ω. With x_n already chosen, where $n \geqslant 1$, let us set

$$x_{n+1} = w_{\alpha_n}(x_n),$$

where $0 < \alpha_n \leqslant \alpha'$ and $w_{\alpha_n}(x_n)$ is an arbitrary element of the set $P_{\alpha_n}(x_n)$.

THEOREM 2.10. *Let* $\{x_n\}$ *denote the sequence constructed according to the procedure described above. Then,*

$$\lim_{n\to\infty} \| x_n - x_{n+1} \| = \lim_{n\to\infty} \| x_n - w_{\alpha_n}(x_n) \| = 0$$

and, furthermore, the sequence $\{f(x_n)\}$ *decreases monotonically.*

Proof: Since the functional f satisfies the Lipschitz condition (1.5) on the convex hull of the set Ω, Lemma 1.2 is applicable. According to it,

$$f(x_{n+1}) = f(w_{\alpha_n}(x_n)) \leqslant f(x_n) + (Fx_n, w_{\alpha_n}(x_n) - x_n) + \\ + \frac{L}{2} \| w_{\alpha_n}(x_n) - x_n \|^2.$$

Using (2.16), we obtain

$$f(x_n) - f(x_{n+1}) \geqslant \left(\frac{1}{2\alpha_n} - \frac{L}{2}\right) \| w_{\alpha_n}(x_n) - x_n \|^2 \geqslant \\ \geqslant \left(\frac{1}{2\alpha'} - \frac{L}{2}\right) \| w_{\alpha_n}(x_n) - x_n \|^2. \tag{2.17}$$

Since

$$\varepsilon = \frac{1}{2\alpha'} - \frac{L}{2} = \frac{1 - L\alpha'}{2\alpha'} > 0$$

(by virtue of the definition of α'), the sequence $\{f(x_n)\}$ decreases monotonically. It follows from our assumptions regarding the operator F that Lemma 1.1 is applicable. Therefore, the functional f is bounded on Ω. It then follows from (2.17) that

$$\lim_{n\to\infty} \|w_{\alpha_n}(x_n) - x_n\|^2 = 0.$$

This completes the proof of the theorem.

3. Minimization of a sublinear functional on a convex compact set.

1°. We now formulate the basic problem of the present chapter.

Problem 1. Let Ω denote a convex compact subset of a normed space X and let p denote a sublinear functional defined on X. Find the element $y \in \Omega$ such that

$$p(y) = \min_{x\in\Omega} p(x).$$

It follows from Theorem 4.1 of Chapter 1 that Problem 1 has a solution. The methods presented below for solving this problem can be regarded as a generalization of the methods proposed in [2, 26, 51]. We note that, just as in Sec. 1, we shall assume known the solution of the problem of minimizing a linear functional on the sets that we are interested in.

2°. Suppose that $g_1, g_2, \ldots, g_n \in X^*$. We set

$$p_n(x) = \max_{k\leqslant n} g_k(x). \tag{3.1}$$

Obviously, the functional p_n defined by formula (3.1) is sublinear. By using Theorem 2.6 of Chapter 1, we can easily show that

$$\mathfrak{u}_{p_n} = \left\{ f \in X^* \mid f = \sum_{k=1}^{n} \gamma_k g_k,\ \gamma_k \geqslant 0,\ k = 1, 2, \ldots, n;\ \sum_{k=1}^{n} \gamma_k = 1 \right\}. \tag{3.2}$$

Let x denote a member of X and suppose that

$$p_n(x) = \max_{k\leqslant n} g_k(x) = g_{k_s}(x)$$

for $s = 1, 2, \ldots, t$. One can easily show that

$$u^x_{p_n}=\left\{f\in X^*\,|\,f=\sum_{s=1}^{t}\gamma_{k_s}g_{k_s},\ \gamma_{k_s}\geqslant 0,\ s=1,\ 2,\ \ldots,\ t,\ \sum_{s=1}^{t}\gamma_{k_s}=1\right\}. \quad (3.3)$$

Let us consider elements $x_1,\ x_2,\ \ldots,\ x_q\in\Omega$. We define

$$\Omega_q=\left\{x\in X\,|\,x=\sum_{i=1}^{q}\alpha_i x_i,\ \alpha_i\geqslant 0,\ i=1,\ 2,\ \ldots,\ q,\ \sum_{i=1}^{q}\alpha_i=1\right\}. \quad (3.4)$$

(Here, Ω_q is a convex polyhedron with vertices at the elements $x_1, x_2, \ldots, x_q$ or, what amounts to the same thing, the convex hull of these elements.) We now formulate the following problem:

Problem 2. Let $x_1, \ldots, x_q$ denote elements of X and let $g_1, \ldots, g_n$ denote elements of X^*. Minimize the functional p_n defined from the functionals $\{g_k\}_{k=1}^{n}$ with the aid of (3.1) over the set Ω_q, the convex hull of the elements $\{x_i\}_{i=1}^{q}$.

Let us show that problem 2 coincides with the basic problem of the theory of matrix games and hence can be solved by the methods of linear programming. We need to find an element

$$x^{(q)}=\sum_{i=1}^{q}\alpha_i' x_i,$$

such that

$$p_n(x^{(q)})=\min_{x\in\Omega_q}p_n(x)=\min_{x\in\Omega_q}\max_{k\leqslant n}g_k(x)=$$
$$=\min_{\alpha}\max_{k\leqslant n}g_k\left(\sum_{i=1}^{q}\alpha_i x_i\right)=\min_{\alpha}\max_{k\leqslant n}\sum_{i=1}^{q}\alpha_i g_k(x_i).$$

Here, the minimum is over all q-tuples $\alpha=(\alpha_1,\ \ldots,\ \alpha_q)$ such that each $\alpha_i\geqslant 0$ for $i=1,\ 2,\ \ldots,\ q$ and

$$\sum_{i=1}^{q}\alpha_i=1.$$

Thus, to solve problem 2, we need to exhibit the q-tuple $(\alpha'=\alpha_1',\ \ldots,\ \alpha_q')$ that provides the minimax, and this is the fundamental problem in the theory of matrix games.

Let us look at the dual problem. Let $\Gamma \subset X^*$ denote the convex hull of the set $\{g_1, g_2, \ldots, g_n\}$:

$$\Gamma = \left\{ f \in X^* \mid f = \sum_{k=1}^{n} \gamma_k g_k,\ \gamma_k \geqslant 0,\ k = 1, \ldots, n;\ \sum_{k=1}^{n} \gamma_k = 1 \right\}.$$

Then, in accordance with the minimax theorem (see [5, Chapter 3]),

$$p_n(x^{(q)}) = \min_{\alpha} \max_{k < n} \sum_{i=1}^{q} \alpha_i g_k(x_i) =$$

$$= \max_{\gamma} \min_{i \leqslant q} \sum_{k=1}^{n} \gamma_k g_k(x_i) = \max_{f \in \Gamma} \min_{i \leqslant q} f(x_i).$$

Let $\gamma' = (\gamma_1', \ldots, \gamma_n')$ denote the n-tuple that provides the maximin. Set

$$f^{(q)} = \sum_{k=1}^{n} \gamma_k' g_k.$$

Let us note certain properties of the functionals $f^{(q)}$.

1) From the minimax theorem,

$$f^{(q)}(x^{(q)}) = \sum_{k=1}^{n} \sum_{i=1}^{p} \alpha_i' \gamma_k' g_k(x_i) = p_n(x^{(q)}).$$

2) It follows from (3.2) that $f^{(q)} \in U_{p_n}$.

3) Since the n-tuple $\gamma' = (\gamma_1', \ldots, \gamma_n')$ provides

$$\max_{\gamma} \min_{i \leqslant q} \sum_{k=1}^{n} \gamma_k g_k(x_i),$$

we have

$$f^{(q)}(x^{(q)}) = p_n(x^{(q)}) = \max_{\gamma} \min_{i \leqslant q} \sum_{k=1}^{n} \gamma_k g_k(x_i) =$$

$$= \min_{i \leqslant q} \sum_{k=1}^{n} \gamma_k' g_k(x_i) = \min_{i \leqslant q} f^{(q)}(x_i),$$

that is, $f^{(q)}(x^{(q)}) \leqslant f^{(q)}(x_i)$ for $i=1, 2, \ldots, q$. Remembering that Ω_q is the convex hull of the elements $\{x_i\}_{i=1}^q$, we have, for arbitrary $x \in \Omega_q$,

$$f^{(q)}(x^{(q)}) \leqslant f^{(q)}(x).$$

It follows from what was said above that

$$f^{(q)} \in u_{p_n}^{x^{(q)}}.$$

Therefore [see (3.3)], $f^{(q)}$ can be represented in the form of a convex combination consisting only of those functionals that provide $\max_{k \leqslant n} g_k(x)$. In particular, if this maximum is attained only on a single functional g_{k_0}, we have $f^{(q)} = g_{k_0}$.

3°. Let us now look at the following more complicated problem:

Problem 3. Let $g_1, g_2, \ldots, g_n$ denote members of X^* and let Ω denote a convex compact subset of X. Minimize on Ω the functional p_n defined with the aid of formula (3.1) from the functionals $\{g_k\}_{k=1}^n$.

To solve problem 3, we can use the following method: We choose arbitrarily an element $\bar{z}_i \in \Omega$ for $i=1, 2, \ldots, r$, where $r \geqslant 2$, and we construct the polyhedron Ω_r, that is, the convex hull of the set $\{\bar{z}_i\}_{i=1}^r$. Solving problem 2, let us find the element z_1 such that

$$p_n(z_1) = \min_{x \in \Omega_r} p_n(x).$$

Solving the dual problem, let us find a functional

$$f_1 = \sum_{k=1}^{n} \gamma_k' g_k \in u_{p_n}^{z_1},$$

such that $f_1(z_1) \leqslant f_1(x)$ for $x \in \Omega_r$. Then let us find an element[1] of the set $G_\Omega f_1$ and let us denote it by z_{r+1} (see Fig. 7).

[1]We recall that $z \in G_\Omega f$ if and only if $f(z) = \min_{x \in \Omega} f(z)$.

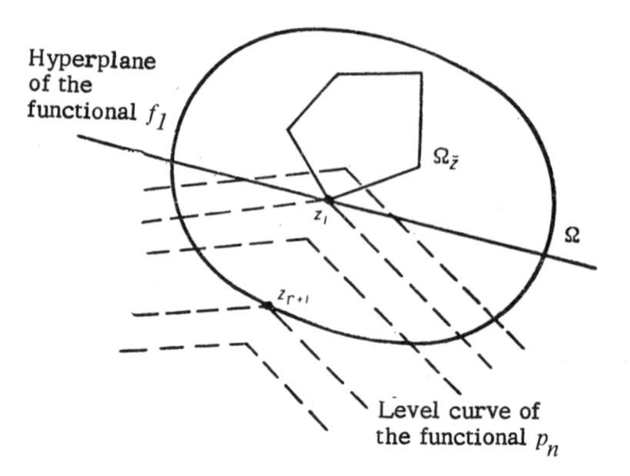

Fig. 7.

Let us now look at the polyhedron Ω_{r+1}, that is, the convex hull of the elements $\bar{z}_1, \bar{z}_2, \ldots, \bar{z}_{r+1}$ and, solving problem 2 and its dual, let us find the element z_2 and the functional f_2. Then, let us find the element $\bar{z}_{r+2} \in G_\Omega f_2$. Then let us construct the polyhedron Ω_{r+2}, etc. In case there exists an m such that

$$f_m(z_m) = f_m(\bar{z}_{r+m}) = \min_{x \in \Omega} f_m(x),$$

it follows from the necessary and sufficient condition for a minimum (II.2.16), since $f_m \in u_{p_n}^{z_m}$, that z_m is the element sought. Otherwise, consider the sequences

$$\begin{array}{l} z_1, \quad z_2, \ldots, z_m, \quad \ldots, \\ \bar{z}_{r+1}, \bar{z}_{r+2}, \ldots, \bar{z}_{r+m}, \ldots, \\ f_1, \quad f_2, \ldots, f_m, \ldots . \end{array}$$

Since $\Omega_m \subset \Omega_{m+1}$, we have $p_n(z_{m+1}) \leqslant p_n(z_m)$.

Let us show that $\{z_m\}$ is a minimizing sequence.

THEOREM 3.1. *The minimum of the functional p_n on the set Ω is attained at the cluster points of the sequence $\{z_m\}$ described above.*

Proof: In the present case, we have for all m

$$f_m(z_m) > f_m(\bar{z}_{r+m}). \tag{3.5}$$

Since Ω is compact, the sequences $\{z_m\}$ and $\{\bar{z}_{r+m}\}$ have cluster points. Since the functionals f_m belong, for all m, to the convex hull of a finite number of functionals $\{g_1, \ldots, g_m\}$, the sequence $\{f_m\}$ also has cluster points. Construct a sequence of numbers $m_1, m_2, \ldots, m_j, \ldots$ such that $\lim z_{m_j}=z$, $\lim z_{r+m_j}=\bar{z}$, and $\lim f_{m_j}=f$. Taking the limit in (3.5), we have

$$f(z)\geqslant f(\bar{z}). \tag{3.6}$$

The functional f_m has the property that

$$f_m(z_m)\leqslant f_m(\Omega_{r+m-1}).$$

Here and in what follows, an inequality of the form $f(x)\leqslant f(\Omega)$ means that $f(x)\leqslant f(y)$ for arbitrary $y\in\Omega$. In particular,

$$f_{m_j}\left(z_{m_j}\right)\leqslant f_{m_j}\left(\Omega_{r+m_j-1}\right).$$

Since

$$\bar{z}_{r+m_j-1}\in\Omega_{r+m_j-1},$$

we have

$$f_{m_j}\left(z_{m_j}\right)\leqslant f_{m_j}\left(\bar{z}_{r+m_j-1}\right),$$

so that $f(z)\leqslant f(\bar{z})$. From this last inequality and inequality (3.6), it follows that $f(z)=f(\bar{z})$. Taking the limit as $m\to\infty$ in the relations $f_m(z_m)=p_n(z_m)$ and $f_m(x)\leqslant p_n(x)$ for $x\in X$, one can easily show that $f\in u_{p_n}^z$. It is also obvious that

$$f(\bar{z})=\min_{x\in\Omega} f(x).$$

It follows from the necessary and sufficient condition for a minimum (II.2.16) that the minimum is attained at the point z. This completes the proof of the theorem.

Remark 1. A minimum is attained at every cluster point of the sequence $\{z_m\}$. If the functional p_m attains a minimum on Ω at a unique point, the sequence $\{z_m\}$ converges.

Remark 2. To get a two-sided estimate of the convergence, let us set

$$\mu = \min_{x \in \Omega} p_n(x) = p_n(z).$$

Then,

$$f_m(\bar{z}_{r+m}) \leqslant \mu \leqslant f_m(z_m). \tag{3.7}$$

The right-hand member of this inequality follows from the fact that

$$f_m(z_m) = p_n(z_m) \geqslant \mu,$$

the left-hand member from the relation

$$f_m(\bar{z}_{r+m}) \leqslant f_m(z) \leqslant p_n(z) = \mu.$$

4°. The solution of problem 1 can be reduced to solving a sequence of problems of the type of problem 3. Let us describe the method proposed.

1) We choose arbitrary elements y_k of Ω and find functionals $g_k \in u_p^{y_k}$ for $k=1, 2, \ldots, r$, where $r \geqslant 2$.

2) Suppose $y_1, y_2, \ldots, y_n$, where $n \geqslant r$, and functionals $g_k \in u_p^{y_k}$ for $k=1, 2, \ldots, n$ have been defined. Consider the functional p_n defined by

$$p_n(x) = \max_{k \leqslant n} g_k(x).$$

Let us solve problem 3 and find the element y_{n+1} minimizing the functional p_n on Ω. Then, let us find the functional $g_{n+1} \in u_p^{y_{n+1}}$. If $y_n = y_{n+1}$, the process terminates. (As we shall show below, in this case, y_n is the element we are seeking.) Otherwise, consider the sequence $y_1, y_2, \ldots, y_n, \ldots$. Let us point out certain properties of this sequence and of the functionals p_n.

$$1)\ p_n(x) \leqslant p_{n+1}(x) \leqslant p(x). \tag{3.8}$$

To see this, note that, since $g_k \in u_p$ for $k=1, \ldots, n+1$, we have

$$p_n(x) = \max_{k<n} g_k(x) \leqslant \max_{k \leqslant n+1} g_k(x) = p_{n+1}(x) \leqslant p(x).$$

2) Suppose that $\mu_n = p(y_n)$. Since $g_k(y_n) \leqslant p(y_n)$ for $k = 1, 2, \dots, n$ with equality holding for the functional g_n, we have

$$\mu_n = p(y_n) = g_n(y_n) = \max_{k \leqslant n} g_k(y_n) = p_n(y_n). \tag{3.9}$$

Obviously,

$$\mu_n = p_{n+q}(y_n) \quad (q = 1, 2, \dots).$$

Define $\lambda_n = p_{n-1}(y_n)$. Let us show that the sequence $\{\lambda_n\}$ is nondecreasing. We have

$$\lambda_n = p_{n-1}(y_n) = \min_{x \in \Omega} p_{n-1}(x) = \min_{x \in \Omega} \max_{k \leqslant n-1} g_k(x) \leqslant$$
$$\leqslant \min_{x \in \Omega} \max_{k < n} g_k(x) = p_n(y_{n+1}) = \lambda_{n+1}.$$

Suppose now that

$$\mu = \min_{x \in \Omega} p(x).$$

LEMMA 3.1. For $n = 1, 2, \dots$

$$\lambda_n \leqslant \mu \leqslant \mu_n. \tag{3.10}$$

Proof: Suppose that an element u attains a minimum p on Ω, that is, $p(u) = \mu$. Since $y_n \in \Omega$, we have $p(y_n) = \mu_n \geqslant \mu$. On the other hand,

$$\mu = p(u) \geqslant \max_{k \leqslant n-1} g_k(u) \geqslant \min_{x \in \Omega} \max_{k \leqslant n-1} g_k(x) = p_{n-1}(y_n) = \lambda_n.$$

THEOREM 3.2. If, for some n, we have $y_n = y_{n+1}$, then the element y_n is a solution of problem 1. In the opposite case, the minimum sought is attained at the cluster points of the sequence $\{y_n\}$.

Proof: It follows from (3.10) that, if $\lambda_n = \mu_n$, then y_n is the element sought. In particular, y_n provides a minimum if $y_n = y_{n+1}$. Indeed, in this case we may assume that $g_n = g_{n+1}$ and, hence,

$$\lambda_{n+1} = p_n(y_{n+1}) = \max_{k \leqslant n} g_k(y_{n+1}) =$$
$$= \max_{k \leqslant n+1} g_k(y_{n+1}) = g_{k+1}(y_{n+1}) = p(y_{n+1}) = \mu_{n+1}.$$

Suppose now that $\lambda_n < \mu_n$, where $n = 1, 2, \ldots$. Let $\{y_{n_j}\}$ denote a convergent subsequence of the sequence $\{y_n\}$ such that $y = \lim y_{n_j}$. Let us define

$$\mu' = p(y);$$

$$\lambda = \lim_{n \to \infty} \lambda_n.$$

(This limit exists since the sequence $\{\lambda_n\}$ is nondecreasing and bounded.) By using (3.10), we obtain

$$\lambda \leqslant \mu = \min_{x \in \Omega} p(x) \leqslant p(y) = \mu'. \tag{3.11}$$

On the other hand, by applying (3.8) and (3.9), we obtain

$$\begin{aligned}\mu_{n_j-1} &= p_{n_j-1}(y_{n_j-1}) = p_{n_j-1}(y_{n_j-1} + y_{n_j} - y_{n_j}) \leqslant \\ &\leqslant p_{n_j-1}(y_{n_j}) + p_{n_j-1}(y_{n_j-1} - y_{n_j}) \leqslant p_{n_j-1}(y_{n_j}) + \\ &+ p_{n_j-1}(y_{n_j} - y_{n_j-1}) = \lambda_{n_j} + p_{n_j-1}(y_{n_j} - y_{n_j-1}).\end{aligned}$$

By taking the limit, we obtain $\mu' \leqslant \lambda$. This last inequality, together with (3.11), yields $\lambda = \mu = \mu'$, from which it follows that y is the element we are seeking.

Remark 1. A minimum is attained at every cluster point of the sequence $\{y_n\}$. If p attains a minimum Ω at a unique point, then the sequence $\{y_n\}$ converges.

Remark 2. $\mu_n \to \mu$.

Remark 3. Inequalities (3.10) provide a convenient two-sided estimate of the convergence.

5°. The method proposed in 3° might be called a method of approximation of a set since its central feature consists in approximating the set Ω in which we are interested close to the point in which we are interested in a special way by means of the polyhedrons constructed. The method proposed in 4° might be called the method of approximation of a functional since we are approximating the sublinear functional p in question with functionals of the form p_n. As was shown above, to solve problem 1, we can apply the method of approximation of a functional but, when we do, we need at every step to solve problem 3 and for this we need to apply the infinite process of the method of approximation of a set. If X is a separable space, we can prove the convergence of the method of simultaneous approximation of a set and a functional to be described below. When using this method, we need to solve at every step only a single problem of the type of problem 2. Here, however, estimates of the type of (3.7) and (3.10) have not been obtained. The method in question is described as follows:

1) We choose arbitrarily r elements $\bar{z}_1, \bar{z}_2, \ldots, \bar{z}_r$ (where $r \geqslant 2$ and $\bar{z}_i \in \Omega$ for $i=1, \ldots, r$ and s functionals $g_1, g_2, \ldots, g_s$ (where $s \geqslant 2$ and $g_k \in \boldsymbol{u}_p$ for $k=1, 2, \ldots, s$).

2) Let $\bar{z}_1, \ldots, \bar{z}_{r+m}$ denote elements and let $g_1, g_2, \ldots, g_{s+m}$ (for $m=0, 1, 2, \ldots$) denote functionals already chosen or determined. Consider the functional p_{s+m} defined by

$$p_{s+m}(x) = \max_{k \leqslant s+m} g_k(x)$$

and the polyhedron Ω_{r+m}, which is the convex hull of the elements $\{\bar{z}_i\}_{i=1}^{r+m}$. If we solve problem 2 and the dual problem to it, we get an element z_{m+1} minimizing p_{s+m} on Ω_{r+m} and a functional

$$f_{m+1} \in \boldsymbol{u}_{p_{s+m}}^{z_{m+1}},$$

such that

$$f_{m+1}(z_{m+1}) \leqslant f_{m+1}(\Omega_{r+m}).$$

Let us now choose

$$g_{s+m+1} \in \boldsymbol{u}_p^{z_{m+1}} \text{ and } \bar{z}_{r+m+1} \in G_\Omega f_{m+1}.$$

As a result of all this, we have constructed the sequences

$$\begin{array}{llll} z_1, & z_2, & \ldots, & z_m, \ldots, \\ \bar{z}_{r+1}, & \bar{z}_{r+2}, & \ldots, & \bar{z}_{r+m}, \ldots, \\ f_1, & f_2, \ldots, & & f_m, \ldots, \\ g_{s+1}, & g_{s+2}, & \ldots, & g_{s+m}, \ldots. \end{array}$$

THEOREM 3.3. *The cluster points of the sequence $\{z_m\}$ constitute the solution of problem 1.*

Proof: Since $\bar{z}_{r+m} \in G_\Omega f_m$ we have

$$f_m(z_m) \geqslant f_m(\bar{z}_{r+m}). \tag{3.12}$$

Since $f_m \in U^{z_m}_{p_{s+m-1}}$, it follows that f_m belongs to the convex hull of the functionals $g_1, \ldots, g_{s+m-1}$. Since $g_k \in U_p$ for $k = 1, \ldots, s+m-1$ and the set U_p is convex, it follows that f_m also belongs to U_p. It follows from the separability of X that every bounded weakly sequentially closed set in X^* is weakly compact. In particular, U_p is weakly compact and hence the sequence $\{f_m\}$ has cluster points (in the sense of weak convergence). Let us now choose a sequence of numbers $m_1, m_2, \ldots, m_j, \ldots$ such that the limits

$$\lim z_{m_j} = z, \ \lim \bar{z}_{r+m_j} = \bar{z}, \ \lim f_{m_j} = f$$

exist, the last of these to be understood in the weak sense. Taking the limit in (3.12), we have $f(z) \geqslant f(\bar{z})$. The converse assertion is proven the same way as in the proof of Theorem 3.1. To show that

$$p(z) = \min_{x \in \Omega} p(x),$$

we need only show that, by virtue of (II.2.16), we have $f \in U^z_p$ and $\bar{z} \in G_\Omega f$. One can easily prove this last relationship by taking the limit in the inequality $f_m(\bar{z}_{r+m}) \leqslant f_m(\Omega)$. Also f is obviously a member of U_p. Thus, to complete the proof, we need only to verify that $f(z) \leqslant p(z)$. The inequality $f(z) = p(z)$ follows from the fact that f is a supporting functional. The opposite (conditional) inequality can be obtained by the same line of reasoning as in Theorem 3.2. This completes the proof of the theorem.

6°. Now, let Ω denote an arbitrary (in general, nonconvex) compact set. Let us suppose that we know the solution of the problem of minimizing a functional of the type (3.1) on Ω. Then, to minimize a sublinear functional p on Ω we can use the method of approximation of a functional that was described in 4°. We did not use the convexity of Ω in the proof of either Lemma 3.1 or Theorem 3.2. Thus, the method of approximation of a functional leads to an absolute minimum in the case of a nonconvex as well as a convex compact set.

7°. Let us look at the following problem:

Problem 4. Let X and Y denote normed spaces, let A denote a complete continuous operator mapping X into Y, and let Ω denote a weakly compact subset of X. Let p denote a sublinear functional defined on Y and suppose that p generates on X a functional J defined by

$$J(x)=p(Ax)$$

for $x \in X$. Minimize J on Ω.

To solve problem 4, let us make the following modification in the method of approximation of a functional.

1) Let us choose arbitrarily elements $x_k \in \Omega$ for $k=1, \ldots, r$, where $r \geqslant 2$. Define $y_k=Ax_k$. Let us find the functionals

$$g_k \in U_p^{y_k}$$

for $k=1, \ldots, r$.

2) Suppose that we have either chosen or constructed the elements $x_1, \ldots, x_r, \ldots, x_n$ and functionals $g_k \in U_p^{y_k}$, where $y_k=Ax_k$ for $k=1, \ldots, n$. Consider functional J_n defined by

$$J_n(x)=\max_{k \leqslant n} g_k(Ax).$$

By minimizing J_n on Ω (we assume that the solution of this problem is known), let us find the element x_{n+1} such that

$$J_n(x_{n+1})=\min_{x \in \Omega} J_n(x).$$

Then, let us find the functional $g_{n+1} \in U_p^{y_{n+1}}$, where $y_{n+1}=Ax_{n+1}$. Reasoning as in the proof of Theorem 3.2, we can easily prove

THEOREM 3.4. *If $x_n=x_{n+1}$ for some n, then the element x_n is a solution of Problem 4. Otherwise, the maximum that we are seeking is attained at the cluster points (in the sense of weak convergence) of the sequence $\{x_n\}$.*

Let us give an example. Suppose that Ω is a closed bounded subset of R^n and that E is a closed bounded subset of R^m. Let $f(x, y)$ denote a function defined on the set

$$\Omega \times E=\{(x, y) \mid x \in \Omega, y \in E\}.$$

Suppose that f is continuous with respect to y for every fixed x in Ω and that it satisfies a Lipschitz condition with respect to x, that is, that there exists a constant $L > 0$ such that, for all x' and x'' in Ω and all y in E,

$$|f(x', y) - f(x'', y)| \leqslant L \|x' - x''\|_{R^n}. \tag{3.13}$$

Find the points $x^0 \in \Omega$ and $y^0 \in E$ such that

$$f(x^0, y^0) = \min_{x \in \Omega} \max_{y \in E} f(x, y).$$

Obviously, to solve this problem, it will be sufficient to find an element $x^0 \in \Omega$ such that

$$\max_{y \in E} f(x^0, y) = \min_{x \in \Omega} \max_{y \in E} f(x, y) \tag{3.14}$$

because if x^0 is known, y^0 can be found from the condition

$$f(x^0, y^0) = \max_{y \in E} f(x^0, y)).$$

Let us define an operator A that maps Ω into $C(E)$ as follows: for $x' \in \Omega$,

$$(Ax')(y) = f(x', y).$$

It follows from (3.13) that A is a completely continuous operator. The problem of finding an element x^0 satisfying (3.14) thus reduces to minimizing on Ω the functional J defined by

$$J(x) = \max_{y \in E} (Ax)(y),$$

in terms of the completely continuous operator A and the sublinear functional p, which is in turn defined by

$$p(u) = \max_{y \in E} u(y)$$

for $u \in C(E)$.

The modification described above in the method of approximation of a functional amounts in the present case to the following procedure:

1) Let us choose arbitrarily elements x_k in Ω for $k=1, \ldots, r$, where $r \geqslant 2$, and let us find points $y_k \in E$ such that

$$f(x_k, y_k) = \max_{y \in E} f(x_k, y).$$

2) Suppose that we have chosen or constructed the elements $x_1, \ldots, x_r, \ldots, x_n$, where $n \geqslant r$, and points $y_1, \ldots, y_n$. The element x_{n+1} is determined from the condition

$$\min_{x \in \Omega} \max_{k \leqslant n} f(x, y_k) = \max_{k \leqslant n} f(x_{n+1}, y_k).$$

Then, we find the point $y_{n+1} \in E$ such that

$$f(x_{n+1}, y_{n+1}) = \max_{y \in E} f(x_{n+1}, y).$$

8°. In conclusion, we note that the method of approximation of a functional has a very significant defect: the dimensionality of the auxiliary problem (problem 3) increases at each step. It would be interesting to ascertain whether it is possible, when solving certain specific problems, to find the $(n+1)$st approximation y_{n+1} by using not all the functionals g_k $(k=1, \ldots, n)$, but only certain of these, independently of the value of n.

CHAPTER 4

Solution of Certain Optimal-Control Problems

1. Statement of the problem.

1°. Numerous extremal problems that are encountered in practice are of the following type:

Let Z and X denote normed spaces, let A denote a continuous operator mapping Z into X, let U denote a subset of Z, let Ω' denote a subset of X, and let U' denote the set

$$U' = A^{-1}(\Omega') \equiv \{u \in Z \mid Au \in \Omega'\}.$$

Let f denote a continuous function defined on the space X and suppose that f generates on Z a functional J defined according to the formula for $u \in Z$:

$$J(u) \doteq f(Au).$$

Minimize J on the set $U \cap U'$. This scheme includes, in particular, certain optimum control problems. In these problems, Z and X are normed spaces of real vector-valued functions defined, as a rule, on some interval of the real axis and the operator A is defined by means of a system of ordinary differential equations. The spaces Z and X are called respectively the ***control space*** and the ***phase space*** or ***trajectory space***. The set U is called the ***class of*** *admissible controls*, and the set Ω' is called the *restriction* to the phase coordinates.

In this chapter, the results of Chapters 2 and 3 are used to investigate and solve certain optimum control problems. We shall assume that Z and X are respectively Hilbert spaces of r-dimensional and n-dimensional real vector-valued functions that are defined and square-summable on the interval $[0, T]$, where T is a fixed number in the interval $0 < T < \infty$. The scalar product in T and X is defined in a natural manner. Thus, if

$$u = (u_1(t), \ldots, u_r(t)) \in Z, \quad v = (v_1(t), \ldots, v_r(t)) \in Z,$$

then

$$(u, v) = \int_0^T \sum_{i=1}^r u_i(t) v_i(t)\, dt.$$

We shall also assume that $\Omega' = X$ and, consequently, $U = Z$ (there are no restrictions on the phase coordinates). Problems with restrictions on the phase coordinates are considered, for example, in [21].

Below, we shall describe some classes of admissible controls U, systems of differential equations that define the operator A, and functionals J which we shall consider.

We note that similar problems were considered in [3, 9, 16-19, 29, 30, 34, 35, 38, 39, 41, 42, 45, 46].

2°. Let us describe admissible classes of controls U. The controlling r-dimensional vector-valued function $u(t) = (u^1(t), \ldots, u^r(t)$ is square-summable $[0, T]$ and it satisfies one of the following restrictions:

$$1) \ |u^i(t)| \leqslant \alpha_i(t), \ i = 1, \ldots, r; \ \ t \in [0, T], \tag{1.1}$$

where the $\alpha_i(t)$ [for $i = 1, \ldots, r$] are nonnegative measurable bounded functions on $[0, T]$;

$$2) \ u^*(t) N(t) u(t) \leqslant \beta(t), \ t \in [0, T], \tag{1.2}$$

where $\beta(t) \geqslant 0$ is a measurable bounded function on $[0, T]$ and $N(t)$ is a symmetric $r \times r$ matrix that is positive-definite on $[0, T]$ with measurable bounded elements on $[0, T]$, the asterisk denoting the transpose.

Special cases of (1.2):

$$2')\ u^*(t)\,Nu(t) \leqslant 1,\ t \in [0, T], \tag{1.3}$$

where N is a real symmetric positive-definite $r \times r$ matrix;

$$2'')\ \sum_{i=1}^{r} (u^i)^2(t) \leqslant 1,\ t \in [0, T]; \tag{1.4}$$

$$3)\ |u^i(t)| = 1;\ i = 1, \ldots, r,\ t \in [0, T]; \tag{1.5}$$

$$4)\ |u^i(t)| \in \{a_{i1}, a_{i2}, \ldots, a_{ip_i}\},\quad t \in [0, T];\ p_i \leqslant P < \infty;\ i = 1, \ldots, r, \tag{1.6}$$

where the a_{ij} are given finite nonnegative numbers;

$$5)\ \int_0^T u^{i2}(t)\,dt \leqslant C_i,\quad i = 1, \ldots, r;\ 0 < C_i < \infty; \tag{1.7}$$

$$6)\ \int_0^T u^*(t)\,N(t)\,u(t)\,dt \leqslant C;\ 0 < C < \infty, \tag{1.8}$$

where $N(t)$ is a symmetric $r \times r$ matrix that is positive-definite on the interval $[0, T]$ and that has square-summable elements on $[0, T]$;

$$6')\ \int_0^T \sum_{i=1}^{r} u^{i2}(t)\,dt \leqslant 1; \tag{1.9}$$

7) The control function $u(t)$ satisfies simultaneously the two conditions

$$\left.\begin{aligned} &|u^i(t)| \leqslant 1,\ t \in [0, T]\ (i = 1, \ldots, r), \\ &\int_0^T u^{i2}(t)\,dt \leqslant C_i,\ 0 < C_i < \infty\ (i = 1, \ldots, r). \end{aligned}\right\} \tag{1.10}$$

We shall denote the classes of controls that satisfy one of the restrictions (1.1)-(1.10) respectively by $U_1 - U_{10}$. We note that the classes $U_1 - U_4$ and $U_7 - U_{10}$ are convex, bounded, and weakly closed. The classes U_5 and U_6 are not convex although they are bounded.

3°. Let us describe the systems of differential equations that we shall consider.

$$\text{I.}\quad \frac{dX(t)}{dt} \equiv \dot{X}(t) = A(t)X(t) + \sum_{i=1}^{r} B_i(t)u^i(t) + F(t), \tag{1.11}$$

$$X(0) = X_0, \tag{1.12}$$

where $A(t)$ is an $n \times n$ matrix and $X(t)$, $F(t)$, and the $B_i(t)$ are n-dimensional vectors ($i = 1, \ldots, r$). The elements of the matrix A and the components of the vectors $F(t)$ and $B_i(t)$ [for $i = 1, \ldots, r$] are assumed to be real, piecewise-continuous, and bounded functions on $[0, T]$.

$$\text{II.}\quad \dot{X}(t) = f(X(t), u(t), t), \tag{1.13}$$

$$X(0) = X_0, \tag{1.14}$$

where $X(t) = (x^1(t), \ldots, x^n(t))$ is an n-dimensional vector-valued function, $u(t) = (u^1(t), \ldots, u^r(t))$ is an r-dimensional vector-valued control function, and $f = (f^1, \ldots, f^n)$ is an n-admissible vector-valued function defined for all $X \in R^n$, $u \in R^r$, and $t \in [0, T]$ that is continuous with respect to t on $[0, T]$ and that has partial derivatives with respect to X and u that satisfy a Lipschitz condition and are bounded for all X, u, and t.

$$\text{III.}\quad \dot{X}(t) = f(X(t), X(t - h_1), u(t), t), \tag{1.15}$$

$$X(t) = X_0(t) \text{ for } t \in [-h_1, 0],\ 0 < h_1 < \infty. \tag{1.16}$$

Here, $X(t)$ and $u(t)$ are respectively n-dimensional and r-dimensional vector-valued functions, $f(X, y, u, t)$ is an n-dimensional vector-valued function defined for all $X \in R^n$, $y \in R^n$, $u \in R^r$, and $t \in [0, T]$ that is continuous with respect to t on $[0, T]$ and that has partial derivatives with respect to X, y, and u that satisfy a Lipschitz condition are founded for all X, y, u, and t, and $X_0(t)$ is a vector-valued initial function defined on $[-h_1, 0]$.

$$\text{IV.}\quad \dot{X}(t) = f(X(t), X(t - h_1(t)), u(t), t), \tag{1.17}$$

$$X(t) = X_0(t) \text{ for } t \in [-h_1(0), 0]. \tag{1.18}$$

Here, the function $\nu(t) = t - h_1(t)$ is a strictly increasing continuously differentiable function on $[0, T]$, and

$$\min_{t\in[0,\,T]} h_1(t) > 0.$$

(Under these assumptions, the inverse function $t = r_1(\nu)$ exists and is strictly increasing and continuously differentiable on the interval $[-h_1(0),\ T - h_1(T)]$. The vector-valued functions $X(t)$, $u(t)$, and $f(X,\ y,\ u,\ t)$ are the same as in the system III. The function $X_0(t)$ is a vector-valued function function defined and continuous on $[-h_1(0),\ 0]$.

Obviously, the systems (1.11), (1.13), and (1.15) are special cases of the system (1.17). We shall denote by $X(t,\ u)$ the solutions of the systems (1.11), (1.13), (1.15), and (1.17) for a chosen u. Under our assumptions regarding the systems of differential equations, the solution $X(t,\ u)$ exists and is unique. For the system (1.11), the solution $X(t,\ u)$ is given by Cauchy's formula

$$X(t) = Y(t)\,X_0 + \int_0^t \sum_{i=1}^{r} Y(t)\,Y^{-1}(\tau)\,B_i(\tau)\,u^i(\tau)\,d\tau + \\ + \int_0^t Y(t)\,Y^{-1}(\tau)\,F(\tau)\,d\tau, \tag{1.19}$$

where $Y(t)$ is the fundamental matrix of the homogeneous part of the system (1.11); that is,

$$\dot{Y}(t) = A(t)\,Y(t), \tag{1.20}$$
$$Y(0) = E. \tag{1.21}$$

A solution $X(t,\ u)$ of the systems (1.13) satisfies the integral equation

$$X(t) = X_0 + \int_0^t f(X(\tau),\ u(\tau),\ \tau)\,d\tau, \tag{1.22}$$

and the solution $X(t,\ u)$ of the systems (1.15) and (1.17) satisfies respectively the integral equations

$$X(t) = X_0 + \int_0^t f(X(\tau),\ X(\tau - h_1),\ u(\tau),\ \tau)\,d\tau, \tag{1.23}$$

$$X(t) = X_0 + \int_0^t f(X(\tau),\ X(\tau - h_1(\tau)),\ u(\tau),\ \tau)\,d\tau, \tag{1.24}$$

where $X_0 = X_0(0)$.

We shall apply the term *set of admissibility* $R(t)$ of the systems (1.11), (1.13), (1.15), and (1.17) at the instant $t \in [0, T]$ for the class of controls U to the set of all $z \in R(t)$ for which there exists a $u \in U$ such that

$$X(t, u) = z.$$

For the system (1.11) at an arbitrary instant t the set of admissibility for the class U_p (where $p=1, \ldots, 4, 7, \ldots, 10$) is a convex closed bounded set in R^n.

4°. On the solutions of systems I-IV treated in subsection 3°, we shall study the following functionals:

$$1) \quad J_1^s(u) = F(X(T, u)). \tag{1.25}$$

Here and in what follows, the superscript s ranges over the values I-IV and indicates which of the systems I-IV is used to determine the vector-valued function $X(t, u)$ that appears in the analytic representation of the functional.

The scalar function F appearing in (1.25) is defined and differentiable on R^n. Also, its gradient satisfies a Lipschitz condition on every bounded subset of R^n.

$$2) \quad J_2^s(u) = \int_0^T g(X(t, u), u(t), t)\, dt \tag{1.26}$$

$$(s = \mathrm{I, II, III, IV}).$$

Here, $g(X, u, t)$ is a scalar function defined for all $X \in R^n$, $u \in R^r$, and $t \in [0, T]$ that is continuous with respect to t on $[0, T]$ and has partial derivatives with respect to X and u satisfying a Lipschitz condition on every set that is bounded with respect to X.

$$3) \quad J_3^s(u) = \int_0^T g(X(t, u), u(t), t)\, dt + F(X(T, u)) \tag{1.27}$$

$$(s = \mathrm{I, II, III, IV}).$$

Here, the functions $g(X, u, t)$ and $F(x)$ are as in (1.26) and (1.25) respectively. Obviously, $J_3 = J_1 + J_2$.

$$4)\ J_4^s(u) = \int_0^T g(X(t,\ u),\ X(t - h_2(t),\ u),\ u(t),\ t)\,dt \qquad (1.28)$$

$$(s = \text{I, II, III, IV}).$$

Here, $\nu(t) = t - h_2(t)$ is a strictly increasing continuously differentiable function on $[0,\ T]$, and $h_2(t) \geqslant 0$ for $t \in [0,\ T]$. Obviously, there exists a function $t = r_2(\nu)$ inverse to $\nu(t)$ that is strictly increasing and continuously differentiable on $[-h_2(0),\ T - h_2(T)]$; $g(X,\ y,\ u,\ t)$ is a scalar-valued function defined for all $X \in R^n$, $y \in R^n$, $u \in R^r$, and $t \in [0,\ T]$ that is continuous with respect to t on $[0,\ T]$ and has partial derivatives with respect to X, y, and u satisfying a Lipschitz condition on every set that is bounded with respect to X and y.

For systems III and IV, we shall assume that, if $h_2(0) > h_1(0)$, then $X(t)$ is defined and continuous on $[-h_2(0),\ -h_1(0)]$. ($X(t)$ is defined on $[-h_1(0),\ 0]$ by (1.16) and (1.18) respectively.) We have

$$F(X(T,\ u)) = F(X_0) + \int_0^T \left(\frac{\partial F(X(t,\ u))}{\partial X}\right)^* \dot{X}(t,\ u)\,dt, \qquad (1.29)$$

where $\dot{X}(t)$ is defined respectively by the systems I-IV. Since the systems I-III are special cases of the system IV, the functionals $J_1^s - J_3^s$ (for $s =$ I, II, III, IV) and J_4^s (for $s =$ I, II, or III) are special cases of the functional J_4^{IV}.

Let us look at the following problem: Find a control $u \in U_p$ (for $p = 1,\ \ldots,\ 10$) such that

$$J_q^s(u) = \min_{v \in U_p} J_q^s(v) \quad (q = 1,\ 2,\ 3,\ 4;\ s = \text{I, II, III, IV}). \qquad (1.30)$$

A control $u \in U_p$ satisfying (1.30) is called an optimal control.

2. Necessary conditions for optimality and successive-approximation methods.

1°. It can be shown that under the assumptions made in Sec. 1 regarding the right-hand members of the systems (1.11), (1.13), (1.15), and (1.17) and also regarding the functions $F(X)$ in (1.25), $g(X,\ u,\ t)$ in (1.26), and $g(X,\ y,\ u,\ t)$ in (1.28), the functionals $J_1^s - J_4^s$

are differentiable and their gradients satisfy a Lipschitz condition on every bounded set. In other words, for every r-dimensional vector-valued function $u(t)$ that is square-summable on $[0, T]$, there exists an r-dimensional vector-valued function $(G_q^s u)(\tau)$ for $q=1, 2, 3, 4$ and $s=$ I, II, III, IV that is square-summable on $[0, T]$ and that satisfies the relationship

$$J_q^s(u+\alpha u) = J_q^s(u) + \alpha \int_0^T (G_q^s u)^*(\tau)\, v(\tau)\, d\tau + o_{u,v}(\alpha) \tag{2.1}$$

for every $v(\tau)=(v_1(\tau), \ldots, v_r(\tau))$ and $\alpha \geqslant 0$. Also, for every bounded subset U of the space in question there exists a constant L_U such that, for every u_1 and u_2 in U,

$$\| G_q^s u_1 - G_q^s u_2 \| \leqslant L_U \| u_1 - u_2 \|. \tag{2.2}$$

We note that, when we are studying the problem of the minimum of a functional J_q^s on a nonconvex set (for example, U_5 or U_6), it is convenient to put formula (2.1) into a somewhat different form. If we set $u+\alpha v = w$ in (2.1), we have for arbitrary square-summable vector-valued functions u and w

$$J_q^s(w) = J_q^s(u) + \int_0^T (G_q^s u)^*(\tau)(w(\tau) - u(\tau))\, d\tau + o(\| w - u \|)$$
$$(q=1, 2, 3, 4), \tag{2.3}$$

where

$$\frac{o(\alpha)}{\alpha} \xrightarrow[\alpha\to 0]{} 0.$$

2°. From the necessary (and, in the case of a convex functional, sufficient) condition for a minimum (II.2.12), we get

THEOREM 2.1. *For a functional J_q^s (where $q = 1, 2, 3,$ or 4 and $s =$ I, II, III, or IV) to attain a minimum out of all controls of the class U_p (where $p=1, \ldots, 4; 7, \ldots, 10$) with the control $u \in U_p$, it is necessary, and in case J_q^s is convex, sufficient that*

$$\min_{v \in U_p} \int_0^T (G_q^s u)^*(\tau)(v(\tau) - u(\tau))\, d\tau = 0. \tag{2.4}$$

Let us now stop to look at the control classes U_5 and U_6. The sets U_5 and U_6 are sufficiently "dense" that, for every $x \in U_p$ (where $p = 5, 6$), the cones $M_x(U_p)$ consist only of 0. Therefore, the necessary condition (II.2.3) is trivial in this case. Nonetheless, using the definition of these sets, we can show that the necessary condition (2.4) holds for them. More precisely, we have

THEOREM 2.2. *For the functional J_q^s (where $q = 1, 2, 3$ or 4 and $s = I, II, III$, or IV) to attain a minimum with the control $u \in U_p$ (where $p = 5, 6$), it is necessary that equation (2.4) hold.*

Proof: Let us suppose that (2.4) does not hold. Then, there exists a measurable set $\omega \subset [0, T]$ and an admissible control $v \in U_p$ such that mes $\omega > 0$ and

$$\int_{\omega} (G_q^s u)^* (\tau)(v(\tau) - u(\tau))\, d\tau \equiv \int_{\omega} \Phi(\tau, v)\, d\tau = -\rho < 0.$$

Let us construct a sequence of sets $\{\omega_i\}$ (for $i = 1, 2, \ldots$) such that

$$\omega_1 \subset \omega; \quad \omega_{i+1} \subset \omega_i \quad (i \geqslant 1); \quad \text{mes}\, \omega_{i+1} = \frac{1}{2} \text{mes}\, \omega_i;$$
$$\int_{\omega_i} \Phi(\tau, v)\, d\tau \leqslant -\frac{\rho_i}{2}.$$

Such a sequence $\{\omega_i\}$ can always be constructed. Let us show, for example, that it is possible to construct a set ω_1. It is always possible to construct sets ω' and ω'' such that

$$\text{mes}\, \omega' = \text{mes}\, \omega'' = \frac{1}{2} \text{mes}\, \omega; \quad \omega = \omega' \cup \omega''; \quad \omega' \cap \omega'' = \varnothing.$$

Since

$$\int_{\omega} \Phi(\tau, v)\, d\tau \equiv \int_{\omega'} \Phi(\tau, v)\, d\tau + \int_{\omega''} \Phi(\tau, v)\, d\tau = -\rho,$$

we have either

$$\int_{\omega'} \Phi(\tau, v)\, d\tau \leqslant -\frac{1}{2}\rho, \text{ or} \int_{\omega''} \Phi(\tau, v)\, d\tau \leqslant -\frac{1}{2}\rho.$$

If

$$\int_{\omega'} \Phi(\tau, v)\, d\tau \leqslant -\frac{1}{2}\rho,$$

we set $\omega_1 = \omega'$. On the other hand, if

$$\int_{\omega''} \Phi(\tau, v)\, d\tau \leqslant -\frac{1}{2}\rho,$$

we set $\omega_1 = \omega''$.

Let us now construct a sequence of equations $\{v_i(\tau)\}$, where

$$v_i(\tau) = \begin{cases} v(\tau), & \tau \in \omega_i, \\ u(\tau), & \tau \overline{\in} \omega_i. \end{cases}$$

Obviously, $v_i(\tau) \in U_p$ for $i = 1, 2, \ldots$. For U_5,

$$|v_i(\tau) - u(\tau)| = \sqrt{\sum_{j=1}^{r} (v_i^j(\tau) - u^j(\tau))^2} \leqslant 2\sqrt{r} \equiv B.$$

For U_6,

$$|v_i(\tau) - u(\tau)| \leqslant \sqrt{\sum_{j=1}^{r} A_j^2} \equiv B,$$

where

$$A_j = \sup_{k,\, l \in \{1, \ldots, r\}} |a_{jk} - a_{jl}|,$$

so that

$$\|v_i - u\| \leqslant \frac{1}{2^i} B.$$

Since

$$\int_0^T \Phi(\tau, v_i)\, d\tau = \int_{\omega_i} \Phi(\tau, v_i)\, d\tau \leqslant -\frac{\rho}{2^i},$$

it follows from (2.3) that

$$J_q^s(u_i) \leqslant J_q^s(u) - \frac{\rho}{2^i} + o(\|v_i - u\|) = J_q^s(u) - \frac{\rho}{2^i} + o\left(\frac{\beta}{2^i}\right).$$

Then, for sufficiently large i,

$$J_q^s(v_i) \leqslant J_q^s(u) - \frac{\rho}{2^{i+1}} < J_q^s(u),$$

which contradicts the assumed optimality of the control u. This contradiction completes the proof of Theorem 2.2. By using Theorems 2.1 and 2.2 and calculating $(G_4^{\mathrm{IV}} u)(\tau)$ (see [17]), we obtain on the basis of (2.4)

THEOREM 2.3. *For a control $u \in U_p$ to minimize J_4^{IV} it is necessary, and in case J_4^{IV} is convex, sufficient that*

$$\min_{v \in U_p} \int_0^T \sum_{i=1}^r \left[\left(\frac{\partial f_u(\tau)}{\partial u^i}\right)^* \psi_u(\tau) + \frac{\partial g_u(\tau)}{\partial u^i}\right](v^i(\tau) - u^i(\tau))\, d\tau = 0, \tag{2.5}$$

where

$$f_u(\tau) = f(X(\tau, u),\ X(\tau - h_1(\tau), u),\ u(\tau), \tau),$$

$$g_u(\tau) = g(X(\tau, u),\ X(\tau - h_2(\tau), u),\ u(\tau), \tau),$$

$$\frac{\partial \psi_u(\tau)}{\partial \tau} = \begin{cases} -\left(\frac{\partial f_u(\tau)}{\partial X}\right)^* \psi_u(\tau) - \left(\frac{\partial f_u(r_1(\tau))}{\partial y}\right)^* \dot{r}_1(\tau)\, \psi_u(r_1(\tau)) - c(\tau) \\ \qquad \text{for } \tau \in [0, T - h_1(T)], \\ -\left(\frac{\partial f_u(\tau)}{\partial X}\right)^* \psi_u(\tau) - C(\tau) \text{ for } \tau \in [T - h_1(T), T], \end{cases} \tag{2.6}$$

$$\psi_u(T) = 0, \tag{2.7}$$

$$\frac{\partial f}{\partial X} = \begin{pmatrix} \frac{\partial f^1}{\partial x^1}, \ldots, \frac{\partial f^1}{\partial x^n} \\ \cdots\cdots \\ \frac{\partial f^n}{\partial x^1}, \ldots, \frac{\partial f^n}{\partial x^n} \end{pmatrix}, \quad \frac{\partial f}{\partial y} = \begin{pmatrix} \frac{\partial f^1}{\partial y^1}, \ldots, \frac{\partial f^1}{\partial y^n} \\ \cdots\cdots \\ \frac{\partial f^n}{\partial y^1}, \ldots, \frac{\partial f^n}{\partial y^n} \end{pmatrix},$$

$$C(t) = \begin{cases} \frac{\partial g_u(t)}{\partial X} + \frac{\partial g_u(r_2(t))}{\partial y} \dot{r}_2(t) \text{ for } t \in [0, T - h_2(T)], \\ \frac{\partial g_u(t)}{\partial X} \qquad \text{for } t \in [T - h_2(T), T], \end{cases} \tag{2.8}$$

$$\frac{\partial f}{\partial u^i} = \left(\frac{\partial f^1}{\partial u^i}, \ldots, \frac{\partial f^n}{\partial u^i}\right), \quad \frac{\partial g}{\partial X} = \left(\frac{\partial g}{\partial x^1}, \ldots, \frac{\partial g}{\partial x^n}\right),$$

$$\frac{\partial g}{\partial y} = \left(\frac{\partial g}{\partial y^1}, \ldots, \frac{\partial g}{\partial y^n}\right).$$

For the class of controls $U_1 - U_6$, we obtain on the basis of (2.4)

THEOREM 2.4. *For a control* $u \in U_p$ *(where* $p = 1, \ldots, 6$*) to minimize the functional* J_4^{IV}, *it is necessary that*

$$\min_{v(t) \in U_p} \sum_{i=1}^{r} \left[\left(\frac{\partial f_u(t)}{\partial u^i}\right)^* \psi_u(t) + \frac{\partial g_u(t)}{\partial u^i}(v^i(t) - u^i(t))\right] = 0 \qquad (2.9)$$

for almost all $t \in [0, T]$.

For the functional J_2^{IV}, we obtain on the basis of Theorem 2.3

THEOREM 2.5. *For a control* $u \in U_p$ *(where* $p = 1, \ldots, 10$*) to minimize the functional* J_2^{II} *out of all the controls in the class* U_p, *it is necessary that*

$$\min_{v \in U_p} \int_0^T \sum_{i=1}^{r} \left[\left(\frac{\partial f_u(\tau)}{\partial u^i}\right)^* \psi_u(\tau) + \frac{\partial g_u(\tau)}{\partial u^i}\right](v^i(\tau) - u^i(\tau))\, d\tau = 0, \qquad (2.10)$$

where

$$\dot{\psi}_u(\tau) = -\left(\frac{\partial f_u(\tau)}{\partial X}\right)^* \psi_u(\tau) - \frac{\partial g_u(\tau)}{\partial X}, \qquad (2.11)$$

$$\psi_u(T) = 0, \qquad (2.12)$$

$$f_u(\tau) = f(X(\tau, u), u(\tau), \tau), \quad g_u(\tau) = g(X(\tau, u), u(\tau), \tau),$$

$$\frac{\partial g}{\partial X} = \left(\frac{\partial g}{\partial x^1}, \ldots, \frac{\partial g}{\partial x^n}\right),$$

$$\frac{\partial f}{\partial X} = \begin{pmatrix} \frac{\partial f^1}{\partial x^1}, & \ldots, & \frac{\partial f^1}{\partial x^n} \\ \cdot & \cdot\ \cdot\ \cdot & \cdot \\ \frac{\partial f^n}{\partial x^1}, & \ldots, & \frac{\partial f^n}{\partial x^n} \end{pmatrix}. \qquad (2.13)$$

If $h_2(t) \equiv h_2$ in (1.28) and the system of differential equations is of the form (1.15), then, in formula (2.5), we have

$$\frac{d\psi_u(\tau)}{d\tau} = \begin{cases} -\left(\frac{\partial f_u(\tau)}{\partial X}\right)^* \psi_u(\tau) - \left(\frac{\partial f_u(\tau + h_1)}{\partial y}\right)^* \psi_u(\tau + h_1) - C(\tau) \\ \qquad \text{for } \tau \in [0, T - h_1], \\ -\left(\frac{\partial f_u(\tau)}{\partial X}\right)^* \psi_u(\tau) - C(\tau) \\ \qquad \text{for } \tau \in [T - h_1, T] \end{cases} \qquad (2.14)$$

$$\psi_u(T) = 0, \qquad (2.15)$$

where

$$C(\tau)=\begin{cases}\dfrac{\partial g_u(\tau)}{\partial X}+\dfrac{\partial g_u(\tau+h_2)}{\partial y} & \text{for } \tau\in[0,\ T-h_2],\\[2mm] \dfrac{\partial g_u(\tau)}{\partial X} & \text{for } \tau\in[T-h_2,\ T].\end{cases} \tag{2.16}$$

The function $\psi_u(\tau)$ in Theorem 2.5 coincides up to sign with the function $\psi(\tau)$ in [45, pp. 23-25]. For the functional J_2^{II} and the control classes (1.1)-(1.4), the necessary condition (2.10) is a linearization of the maximum principle of L. S. Pontryagin. As Yu. F. Kazarinov has pointed out to the authors, one can show that, for the control classes U_7, U_8, and U_9 and for the functional J_2^{II}, the "maximum principle" is of the form

$$\min_{v\in U_p}\int_0^T [f(X(\tau,\ u),\ v(\tau),\ \tau)\psi_u(\tau)+g(X(\tau,\ u),\ v(\tau),\ \tau)\,d\tau=$$

$$=\int_0^T [f_u^*(\tau)\psi_u(\tau)+g_u(\tau)]\,d\tau, \tag{2.17}$$

where $\psi(\tau)$ satisfies the system of differential equations (2.11) with initial conditions (2.12). One may assume that, for the classes U_7, U_8, and U_9 and for the functional J_4^{IV}, the necessary condition ("the maximum principle")

$$\min_{v\in U_p}\int_0^T [f^*(X_u(\tau),\ y_u(\tau),\ v(\tau),\ \tau)\psi_u(\tau)+g(X_u(\tau),\ y_u(\tau),\ v(\tau),\ \tau)]\,d\tau=$$
$$=\int_0^T [f^*(X_u(\tau),\ y_u(\tau),\ u(\tau),\ \tau)\psi_u(\tau)+g(X_u(\tau),\ y_u(\tau),\ u(\tau),\ \tau)]\,d\tau, \tag{2.18}$$

is satisfied, where $\psi_u(\tau)$ satisfies the system (2.6) with special conditions (2.7). For the functional J_2^{I}, we have[1]

$$\frac{\partial f_u(\tau)}{\partial X}=A(\tau),\ \dot\psi_u(\tau)=-A^*(\tau)\psi_u(\tau)-\frac{\partial g_u(\tau)}{\partial X};\ \psi_u(T)=0.$$

[1]If the function $g(X,\ u)$ is convex with respect to X and with respect to u, then J_2^{I} is a convex functional.

From this we obtain, by going through the appropriate calculations,

THEOREM 2.6. *For a control $u \in U_p$ (for $p = 1, \ldots, 10$) to minimize the functional J_2^{I} out of all controls of the class U_p it is necessary (and if the function $g(X, u)$ is convex with respect to both X and u, both necessary and sufficient) that*

$$\min_{v \in U_p} \int_0^T \sum_{i=1}^r \left\{ [\omega^*(t)]^* [Y^{-1}(\tau) B_1(\tau)] + \frac{\partial g_u(\tau)}{\partial u^i} \right\} (v^i(\tau) - u^i(\tau))\, d\tau = 0, \quad (2.19)$$

where

$$\omega(t) = \int_t^T Y^*(\tau) \frac{\partial g_u(\tau)}{\partial X}\, d\tau, \quad (2.20)$$

$$\dot{Y}(\tau) = A(\tau) Y(\tau), \quad (2.21)$$

$$Y(0) = E. \quad (2.22)$$

3°. We shall refer to a control $u \in U_p$ (for $p = 1, \ldots, 10$) that satisfies (2.5) as a stationary control. Let us look at the methods of successive approximations for finding stationary controls.

To begin with, let us look at methods that can be used for the control classes $U_1 - U_4$ and $U_7 - U_{10}$ (these classes are convex, weakly closed, and bounded). Suppose that $u \in U_p$ (where $p = 1, \ldots, 4$; 7-10). Let us denote by $v[u]$ any control $v(t) \in U_p$ that satisfies the condition

$$\min_{w \in U_p} \int_0^T \sum_{i=1}^r \left[\left(\frac{\partial f_u(\tau)}{\partial u^i} \right)^* \psi_u(\tau) + \frac{\partial g_u(\tau)}{\partial u^i} \right] w^i(\tau)\, d\tau =$$

$$= \int_0^T \sum_{i=1}^r \left[\left(\frac{\partial f_u(\tau)}{\partial u^i} \right)^* \psi_u(\tau) + \frac{\partial g_u(\tau)}{\partial u^i} \right] v^i(\tau)\, d\tau. \quad (2.23)$$

In other words, $v[u]$ is an element of the set $G_{U_p} l$, where the linear functional l is defined by the vector-valued function

$$\left(\left(\frac{\partial f_u(\tau)}{\partial u^1} \right)^* \psi_u(\tau) + \frac{\partial g_u(\tau)}{\partial u^1}, \ldots, \left(\frac{\partial f_u(\tau)}{\partial u^r} \right)^* \psi_u(\tau) + \frac{\partial g_u(\tau)}{\partial u^r} \right).$$

To find $v[u]$, it is necessary to be able to minimize a linear functional. For the classes $U_1 - U_4$, $U_7 - U_{10}$, this problem was solved in subsections 1° and 2° of Sec. 3. To find stationary controls, we can apply the algorithms expounded in Secs. 1 and 2 of Chapter 3. Let us describe some of these in connection with the present case. For the remainder of this section, let us agree, unless the contrary is stated, to denote by the symbol J any one of the functions J_q^s (for $q = 1, 2, 3$, or 4 and $s =$ I, II, III, or IV) and to denote by U any one of the sets U_p (for $p = 1$-4; 7-10).

1. The conditional-gradient method. As our first approximation, we take an arbitrary admissible $u_1 \in U$. Suppose that $u_k \in U$ has been found. We define $v[u_k] = v_k \in U$. We set up a linear combination $u_{k\alpha} = \alpha u_k + (1-\alpha) v_k$ for $\alpha \in [0, 1]$ and we find $\alpha_k \in [0, 1]$ such that

$$J(u_{k\alpha_k}) = \min_{\alpha \in [0, 1]} J(u_{k\alpha}).$$

We set $u_{k+1} = u_{k\alpha_k}$. Obviously, $J(u_{k+1}) \leqslant J(u_k)$. Thus, the sequence

$$u_1, u_2, \ldots; \; u_k \in U \tag{2.24}$$

that we have constructed is such that $J(u_1) \geqslant J(u_2) \geqslant \ldots$. Since the functional $J(u)$ is bounded below on U, there exists

$$\lim_{k\to\infty} J(u_k) = J^* > -\infty, \quad J(u_k) \geqslant J^*.$$

The sequence of controls (2.24) converges to a stationary control in the following sense:

THEOREM 2.7.

$$\lim_{k\to\infty}\left\{\min_{w\in U}\int_0^T \sum_{i=1}^r \left[\left(\frac{\partial f_{u_k}(\tau)}{\partial u^i}\right)^* \psi_{u_k}(\tau) + \frac{\partial g_{u_k}(\tau)}{\partial u^i}\right] \times \right.$$
$$\left. \times \left(w^i(\tau) - u^i(\tau)\right) d\tau \right\} = 0. \tag{2.25}$$

Theorem 2.7 follows from Theorem 1.1 of Chapter 3. We note that, if a functional is of the form

$$J(u)=\int_0^T g(X(t))\,dt, \tag{2.26}$$

then, for the linear system I, we do not need to use the linear combination $u_{k\alpha}$; instead, it will be sufficient to consider the functional (2.26) only on the combination of solutions

$$X_k(t)=X(t,\ u_k),\ X(t,\ v_k)=\overline{X}_k(t)$$

and seek $\alpha_k \in [0,\ 1]$ such that

$$J(u_{k\alpha_k})\equiv\int_0^T g(X_{k\alpha_k}(t))\,dt=\min_{\alpha\in[0,\ 1]}\int_0^T g(X_{k\alpha}(t))\,dt,$$
$$X_{k\alpha}=\alpha X_k+(1-\alpha)\overline{X}_k,$$

that is, it is not necessary to integrate the system of differential equations (1.11) repeatedly.

On the other hand, if the functional $J(u)$ is the terminal functional J_1^{I}, then, to find α_k, it is sufficient to consider the interval $[X(T,\ u_k),\ X(T,\ v_k)]$ and to find $\alpha_k \in [0,\ 1]$ such that

$$g(X_{k\alpha_k})=\min_{\alpha\in[0,\ 1]} g(X_{k\alpha}),$$

where $X_{k\alpha}=\alpha X(T,\ u_k)+(1-\alpha)X(T,\ v_k)$. We note that, to minimize the functional J_1^{I} and the functional (2.26) in the case of a linear system, the use of the conditional-gradient method is exceptionally effective.

2. The gradient projection method. Suppose that $u \in U_p$ (where $p=1,\ \ldots,\ 4;\ 7,\ \ldots,\ 10$). Consider the vector-valued function $u_\alpha(t)= = u(t)-\alpha G_u(t)$, where $\alpha \in [0,\ \infty)$ and

$$G_u(t)=\left(G_u^1(t),\ \ldots,\ G_u^r(t)\right)$$

is the gradient of the functional $J(u)$. Let us find $w_\alpha^u \in U$ such that[1]

[1]In Sec. 2 of Chapter 3, the element w_α^u was denoted by $w_\alpha(u)$.

$$\int_0^T (w_\alpha^u(t) - u_\alpha(t))^2 \, dt = \min_{v \in U} \int_0^T (v(t) - u_\alpha(t))^2 \, dt. \tag{2.27}$$

In subsection 3° of Sec. 3, we considered the question of finding w_α^u. As our first approximation, let us take any $u_1 \in U$. Suppose that we have found u_k. We define

$$u_{k+1} = u_k + \beta_k \left(w_{\alpha_k}^k - u_k\right), \tag{2.28}$$

where

$$w_{\alpha_k}^k \equiv w_{\alpha_k}^{u_k}.$$

The choice of the coefficients α_k and β_k depends on which of the four schemes of the gradient projection method is used (see Sec. 2, Chapter 3). To solve the problem of the optimal control, it is convenient to use these schemes in the following forms:

The first scheme:

$$u_{k+1} = u_k + \beta_k (w_\alpha^k - u_k), \tag{2.29}$$

where $\alpha \in (0, \infty)$, α is independent of k, and β_k is determined from the condition

$$J\left(u_k + \beta_k \left(w_\alpha^k - u_k\right)\right) = \min_{\beta \in [0,1]} J\left(u_k + \beta (w_\alpha^k - u_k)\right).$$

The second scheme:

$$u_{k+1} = w_\alpha^k, \tag{2.30}$$

where $0 < \alpha < \frac{2}{L_U}$, the number L_U being determined for the set U and the functional J in accordance with formula (2.2) [in this scheme, $\beta_k = 1$].

The third scheme:

$$u_{k+1} = w_{\alpha_k}^k, \tag{2.31}$$

where α_k is determined from the condition

$$J(w^k_{\alpha_k}) = \min_{0 \leqslant \alpha < \infty} J(w^k_\alpha). \tag{2.32}$$

The fourth scheme:

$$u_{k+1} = u_k + \beta_k (w^k_{\alpha_k} - u_k), \tag{2.33}$$

where

$$J(u_k + \beta_k (w^k_{\alpha_k} - u_k)) = \min_{\substack{\alpha \in [0, \infty] \\ \beta \in [0, 1]}} J(u_k + \beta (w^k_\alpha - u_k)) .$$

The convergence of all these schemes of the method was proven in Sec. 2 of Chapter 3. To solve our problem, we can also apply other schemes of the method. For example, α_k in formula (2.28) can be chosen from conditions (2.32) and the coefficient β_k from the condition

$$J(u_k + \beta_k (w^k_{\alpha_k} - u_k)) = \min_{\beta \in [0, 1]} J(u_k + \beta (w^k_{\alpha_k} - u_k)).$$

We note that the methods (2.30) and (2.31) can also be applied to the control classes U_5 and U_6. For these classes (and for the classes $U_1 - U_4$), we can apply yet another method. We shall illustrate it with the example of the class U_5.

Suppose that we have found $u_k \in U_5$ and that u_k is not a stationary control. Let us find $\Omega_k \subset [0, T]$, where the necessary condition (2.9) (where the set Ω_k is measurable) is violated. Let us construct a sequence of sets

$$\omega^k_1, \omega^k_2, \ldots ; \omega^k_i \in \Omega_k ; \omega^k_1 \supset \omega^k_2 \supset \ldots ; d^k_i \xrightarrow[i \to \infty]{} 0,$$

where d^k_i is the measure of the set ω^k_i.

Then, there exists an m_k such that, if $i > m_k$, then, for the control

$$u_{k_i}(t) = \begin{cases} -u_k(t), & t \in \omega^k_i, \\ u_k(t), & t \overline{\in} \omega^k_i \end{cases}$$

we have $J(u_{k_i}) < J(u_k)$. Let us set $u_{k+1} = u_{k i_k}$, where

$$J\left(u_{ki_k}\right) = \min_{i > m_k} J(u_{ki}).$$

Obviously, $J(u_{k+1}) < J(u_k)$. From this point, we proceed in an analogous manner. One can show that the sequence of controls $\{u_k\}$ constructed in this way converges in a certain sense to a stationary control.

Remark. Knowing how to solve the problem of minimizing an integral functional on $[0, T]$, we can now solve a number of other problems:

1. Problems involving linear and nonlinear high-speed operation, that is, the least time transition of a system from one point to another.

2. The problem of least time shift from one surface onto another.

3. Solution of auxiliary problems.

Here, we shall solve the problem of minimizing a linear functional on the sets U_p and the problem of projection onto U_p (where $p = 1, \ldots, 10$).

1°. Let us show how to find, for the control classes $U_1 - U_{10}$, the minimum of a linear integral functional of the form

$$l(u) = \int_0^T \sum_{i=1}^r \Phi_i(t)\, u^i(t)\, dt, \tag{3.1}$$

where the function

$$\Phi(t) = (\Phi_1(t), \ldots, \Phi_r(t))$$

is a measurable bounded vector-valued function that is not identically equal to zero.

For the control classes $U_1 - U_9$, the solution of this problem can be obtained by means of the same considerations as in Sec. 5 of Chapter 1.

1) For the class U_1, the minimum of (3.1) is attained with the vector-valued function

$$v(t)=(v^1(t),\ \dots,\ v^r(t)),$$

where

$$v^i(t)=-\alpha_i(t)\,\mathrm{sign}\,\Phi_i(t)\quad (i=1,\ \dots,\ r).$$

2) For the class U_2, the minimum of (3.1) is attained with

$$v(t)=\begin{cases} 0, & \text{if } \Phi(t)=0,\\ -\dfrac{\sqrt{\beta(t)}}{\|\Phi(t)\|_N}N^{-1}(t)\,\Phi(t), & \text{if } \Phi(t)\neq 0,\end{cases}$$

where

$$\|\Phi(t)\|_N^2=\Phi^*(t)\,N^{-1}(t)\,\Phi(t).$$

3) For U_3, the minimum of (3.1) is attained with

$$v(t)=\begin{cases} 0\quad, & \text{if } \Phi(t)=0,\\ -\dfrac{N^{-1}\Phi(t)}{\|\Phi(t)\|_N}, & \text{if } \Phi(t)\neq 0.\end{cases}$$

4) For U_4, the minimum of (3.1) is attained with

$$v(t)=\begin{cases} 0, & \text{if } \Phi(t)=0,\\ -\|\Phi(t)\|^{-1}\Phi(t), & \text{if } \Phi(t)\neq 0.\end{cases}$$

5) For U_5, the minimum of (3.1) is attained with

$$v(t)=(v_1(t),\ \dots,\ v_r(t)),$$

where

$$v_i(t)=\begin{cases} -1, & \Phi_i(t)=1,\\ \ \ 0, & \Phi_i(t)=0\quad (i=1,\ \dots,\ r),\\ \ \ 1, & \Phi_i(t)=-1.\end{cases}$$

6) For U_6, the minimum of (3.1) is attained with

$$v^i(t) = \begin{cases} \min\limits_{j=1,\ldots,p_i} a_{ij}, & \Phi_i(t) > 0, \\ \max\limits_{j=1,\ldots,p_i} a_{ij}, & \Phi_i(t) \leqslant 0 \end{cases} \quad (i=1, \ldots, r).$$

7) For U_7, the minimum of (3.1) is attained with

$$v(t) = (v_1(t), \ldots, v_r(t)),$$

where

$$v^i(t) = -\frac{\sqrt{C_i}\,\Phi_i(t)}{\|\Phi_i\|} \quad (i=1, \ldots, r).$$

Here, it is assumed that for no i is the function $\Phi_i(t)$ identically zero. If $\Phi_i(t) \equiv 0$ for some i, we set, for this i,

$$v^i(t) \equiv 0 \quad (t \in [0, T]).$$

8) For U_8, the minimum of (3.1) is attained with

$$v(t) = -\frac{\sqrt{C}}{\|\Phi\|_N} N^{-1}(t)\,\Phi(t),$$

where

$$\|\Phi\|_N^2 = \int_0^T \Phi^*(t)\,N^{-1}(t)\,\Phi(t)\,dt.$$

9) It follows from this that, for U_9,

$$v(t) = -\frac{\Phi(t)}{\|\Phi\|}.$$

2°. Let us now look at the problem of minimizing the functional (3.1) on the set U_{10}. The minimum of the functional l on this set is attained, as one can easily show, with the vector-valued function $v(t) = (v^1(t), \ldots, v^r(t))$, where $v^i(t)$, for $i=1, \ldots, r$, minimizes the functional

$$l_i\left(l_i(u)=\int_0^T \Phi_i(t)\,u(t)\,dt\right)$$

for bounded $|u(t)|\leqslant 1$ and

$$\int_0^T u^2(t)\,dt\leqslant C.$$

Thus, it will be sufficient for us to know how to solve the following problem:

Suppose that a is a measurable scalar-valued function that is bounded on $[0,\ T]$. Minimize the functional

$$J(u)=\int_0^T a(t)\,u(t)\,dt \tag{3.2}$$

on the set U, where

$$U=\left\{u\in L^2[0,\ T]\,\middle|\,|u(t)|\leqslant 1\ (t\in[0,\ T]),\ \int_0^T u^2(t)\,dt\leqslant C\right\}.$$

Together with our problem, let us look at the problem of minimizing, for $\lambda\geqslant 0$, the functional

$$J_\lambda(u)=\int_0^T\left[a(t)\,u(t)+\frac{\lambda}{2}u^2(t)\right]dt-\frac{\lambda}{2}C \tag{3.3}$$

on the set $U'=\{u\in L^2[0,\ T]\,|\,|u(t)|\leqslant 1\}$.

It is easy to show that, for $\lambda>0$, the minimum of $J_\lambda(u)$ on the set U' is attained with

$$u_\lambda(t)=-\operatorname{sat}\frac{a(t)}{\lambda}, \tag{3.4}$$

where

$$\operatorname{sat} x=\begin{cases} x, & |x|\leqslant 1,\\ \operatorname{sign} x, & |x|>1.\end{cases} \tag{3.5}$$

If $\lambda = 0$, the minimum of $J_\lambda(u)$ and $J(u)$ on U' is attained with

$$u_0(t) = -\operatorname{sign} a(t).$$

Here, if $\int_0^T u_0^2(t)\,dt \leqslant C$, then the control u_0 minimizes the functional J on the set U.

Let us now suppose that $\int_0^T u_0^2(t)\,dt > C$. For $\lambda > 0$, consider the function

$$h(\lambda) = \int_0^T u_\lambda^2(t)\,dt. \tag{3.6}$$

Let us suppose that the equation

$$h(\lambda) - C = 0 \tag{3.7}$$

has a positive root λ'. Then,

$$J_{\lambda'}(u_{\lambda'}) = \int_0^T a(t)\,u_{\lambda'}(t)\,dt + \frac{\lambda'}{2}\int_0^T u_{\lambda'}^2(t)\,dt - \frac{\lambda'}{2}C =$$
$$= \int_0^T a(t)\,u_{\lambda'}(t)\,dt + \frac{\lambda'}{2}(h(\lambda') - C) = J(u_{\lambda'}).$$

Therefore, for $u \in U$,

$$J(u) = \int_0^T a(t)\,u(t)\,dt \geqslant \int_0^T a(t)\,u(t)\,dt +$$
$$+ \frac{\lambda'}{2}\left(\int_0^T u^2(t)\,dt - C\right) = J_{\lambda'}(u) \geqslant \min_{u \in U'} J_{\lambda'}(u) = J_{\lambda'}(u_{\lambda'}) = J(u_{\lambda'}),$$

from which it follows that the element $u_{\lambda'}$ minimizes the functional J on the set U. Thus, in the present case, that is, the case $\int_0^T u_0^2(t)\,dt > C$, our problem reduces to finding a positive root of equation (3.7).

Let us examine the function $h(\lambda)$ in greater detail. It follows from (3.4) and (3.6) that, for $\lambda > 0$,

$$h(\lambda)=\int_0^T\left(\operatorname{sat}\frac{a(t)}{\lambda}\right)^2 dt. \tag{3.8}$$

Let us extend the domain of definition of h to include $\lambda=0$ by setting

$$h(0)=\int_0^T(\operatorname{sign} a(t))^2\,dt.$$

Obviously,

$$h(\lambda)\underset{\lambda\to 0}{\longrightarrow} h(0).$$

One can easily show that the function $h(\lambda)$ is continuous and non-increasing on $[0, \infty)$ and that, since $a(t)$ is bounded,

$$h(\lambda)\underset{\lambda\to\infty}{\longrightarrow} 0.$$

It follows, in particular, that equation (3.7) has a positive solution when

$$h(0)=\int_0^T u_0(t)\,dt > C.$$

For arbitrary real μ, let us define

$$E_\mu=\{t\in[0, T]\,|\,a(t)=\mu\}.$$

Let us show that, if the function a is such that

$$\operatorname{mes} E_\mu=0 \tag{3.9}$$

for arbitrary μ, then the function $h(\lambda)$ is continuously differentiable for $\lambda>0$. Let λ denote a number in $(0, \infty)$. We define

$$\Omega_\lambda=\left\{t\in[0, T]\,\middle|\,\frac{|a(t)|}{\lambda}\leqslant 1\right\}.$$

Let Δ denote a positive number. Obviously, for $t \in \Omega_\lambda$,

$$\left| \frac{a(t)}{\lambda + \Delta} \right| < 1.$$

For given Δ, let us define

$$\Omega_{1\Delta} = \left\{ t \in [0, T] \, \middle| \, \left| \frac{a(t)}{\lambda + \Delta} \right| \geqslant 1 \right\},$$
$$\Omega_{2\Delta} = \left\{ t \in [0, T] \, \middle| \, \left| \frac{a(t)}{\lambda} \right| \in \left(1, 1 + \frac{\Delta}{\lambda}\right) \right\}.$$

If $t \in \Omega_{1\Delta}$, then

$$\left| \frac{a(t)}{\lambda} \right| = \left| \frac{a(t)}{\lambda + \Delta} \right| \frac{\lambda + \Delta}{\lambda} \geqslant 1 + \frac{\Delta}{\lambda}.$$

Therefore, $\Omega_{1\Delta} \cap \Omega_{2\Delta} = \varnothing$. Obviously, we also have $\Omega_\lambda \cap \Omega_{i\Delta} = \varnothing$ (for $i = 1, 2$) and

$$\Omega_\lambda \cup \Omega_{1\Delta} \cup \Omega_{2\Delta} = [0, T].$$

We note that it follows from (3.9) that mes $\Omega_{2\Delta} \to 0$ as $\Delta \to 0$. Let us now calculate

$$\frac{dh(\lambda + 0)}{d\lambda} = \lim_{\Delta \to +0} \frac{h(\lambda + 0) - h(\lambda)}{\Delta}.$$

Suppose that

$$\left(\operatorname{sat} \frac{a(t)}{\lambda + \Delta}\right)^2 - \left(\operatorname{sat} \frac{a(t)}{\lambda}\right)^2 = \Phi_\lambda(t, \Delta).$$

Then,

$$h(\lambda + \Delta) - h(\lambda) = \int_0^T \Phi_\lambda(t, \Delta)\, dt =$$
$$= \int_{\Omega_\lambda} \Phi_\lambda(t, \Delta)\, dt + \int_{\Omega_{1\Delta}} \Phi_\lambda(t, \Delta)\, dt + \int_{\Omega_{2\Delta}} \Phi_\lambda(t, \Delta)\, dt.$$

Obviously, $\Phi_\lambda(t, \Delta) = 0$ for $t \in \Omega_{1\Delta}$. For $t \in \Omega_{2\Delta}$, we have

$$\Phi_\lambda(t,\ \Delta)=\left(\frac{a(t)}{\lambda+\Delta}\right)^2-1\in\left[-\Delta\frac{2\lambda+\Delta}{(\lambda+\Delta)^2},\ 0\right].$$

Therefore, by the theorem of the mean,

$$h(\lambda+\Delta)-h(\lambda)=\int_{\Omega_\lambda}\Phi_\lambda(t,\ \Delta)\,dt-\operatorname{mes}\Omega_{2\Delta}\theta(\Delta)\Delta,$$
$$\theta(\Delta)\in\left[0,\ \frac{2\lambda+\Delta}{(\lambda+\Delta)^2}\right].$$

Remembering that

$$\operatorname{mes}\Omega_{2\Delta}\xrightarrow[\Delta\to+0]{}0,$$

we now see that

$$\frac{dh(\lambda+0)}{d\lambda}=\lim_{\Delta\to+0}\frac{1}{\Delta}\int_{\Omega_\lambda}\Phi_\lambda(t,\ \Delta)\,dt.$$

But, for $t\in\Omega_\lambda$,

$$\Phi_\lambda(t,\ \Delta)=\left(\frac{a(t)}{\lambda+\Delta}\right)^2-\left(\frac{a(t)}{\lambda}\right)^2,$$
$$\int_{\Omega_\lambda}\Phi_\lambda(t,\ \Delta)\,dt=\left[\frac{1}{(\lambda+\Delta)^2}-\frac{1}{\lambda^2}\right]\int_{\Omega_\lambda}a^2(t)\,dt,$$

so that

$$\frac{dh(\lambda+0)}{d\lambda}=-\frac{2}{\lambda^3}\int_{\Omega_\lambda}a^2(t)\,dt. \tag{3.10}$$

Reasoning as above, we can easily show that

$$\frac{dh(\lambda-0)}{d\lambda}=-\frac{2}{\lambda^3}\int_{\Omega'_\lambda}a^2(t)\,dt, \tag{3.11}$$

where

$$\Omega'_\lambda=\left\{t\in[0,\ T]\,\middle|\,\left|\frac{a(t)}{\lambda}\right|<1\right\}.$$

It follows from (3.9)-(3.11) that the function h is continuously differentiable for arbitrary $\lambda > 0$ and

$$\frac{dh}{d\lambda} = -\frac{2}{\lambda^3} \int_{\Omega_\lambda} a^2(t)\, dt.$$

Thus, if (3.9) is satisfied, then, to solve equation (3.7) [and hence to solve the problem of minimizing the functional J on the set U], we can use the familiar methods of finding the roots of a continuously differentiable function.

Remark 1. If $|a(t)| \geqslant A > 0$, then, for $\lambda < A$,

$$\Omega_\lambda = \Omega'_\lambda = \varnothing \text{ and } \frac{dh(\lambda+0)}{d\lambda} = \frac{dh(\lambda-0)}{d\lambda} = 0,\ h(\lambda) = T.$$

If $|a(t)| \leqslant B < \infty$ and $B > 0$, then, for $\lambda > B$,

$$\frac{dh(\lambda+0)}{d\lambda} = \frac{dh(\lambda-0)}{d\lambda} = -\frac{2}{\lambda^3} \int_0^T a^2(t)\, dt, \quad h(\lambda) = \frac{1}{\lambda^2} \int_0^T a^2(t)\, dt.$$

Remark 2. The problem is solved in an analogous way if instead of the restriction $|u(t)| \leqslant 1$, we have the restriction

$$|u(t)| \leqslant \alpha(t), \quad t \in [0, T],$$

where $\alpha(t)$ is a nonnegative function that is measurable and bounded on $[0, T]$. This can be extended to the case in which the restriction

$$\int_0^T u^2(t)\, dt \leqslant C$$

is replaced with the restriction

$$\int_0^T \beta(t)\, u^2(t)\, dt \leqslant C,\ C > 0,$$

where $\beta(t)$ is a positive measurable bounded function on $[0, T]$.

Remark 3. It is of interest to apply the approach described above to the solution of the problem of finding

$$\min_{u \in U} \sum_{i=1}^{r} \int_0^T a_i(t) u^i(t)\, dt,$$

where $u=(u^1, \ldots, u^r) \in U$, the $u^i(t)$ [for $i=1, \ldots, r$] are measurable functions on $[0, T]$,

$$|u^i(t)| \leqslant \alpha_i(t); \quad i=1, \ldots, r; \quad t \in [0, T],$$

$$\int_0^T u^*(t) N(t) u(t)\, dt \leqslant C; \quad C > 0,$$

the $\alpha_i(t) \geqslant 0$ are measurable bounded functions on $[0, T]$, and $N(t)$ is an $r \times r$ matrix with measurable bounded elements on $[0, T]$.

Remark 4. In an analogous manner, we can solve the following problem, which is the discrete analog of the fundamental problem considered above: find

$$\min \sum_{i=1}^{N} a_i u^i,$$

where $u^i \in [\alpha_i, \beta_i]$ (for $i=1, \ldots, N$), and

$$\sum_{i=1}^{N} u^{i2} \leqslant C,$$

where $C>0$, the α_i and β_i are finite real numbers. In this case, the function $h(\lambda)$ is piecewise differentiable.

3°. To apply the gradient projection method, we need to know how to find the projection of a measurable bounded vector-valued function $G(t)=(g_1(t), \ldots, g_r(t))$ onto the set U. For the sets U_2, U_3, and U_8, this problem is rather laborious. Therefore, the application of projective methods is tedious. For the remaining sets, the problem of finding the projection does not present any difficulty. Thus, it is necessary to find a vector-valued function $v \in U$ such that

$$\min_{u \in U} \int_0^T (u(t) - G(t))^2 \, dt = \int_0^T (v(t) - G(t))^2 \, dt.$$

1) For the set U_1,

$$v^i(t) = \begin{cases} g_i(t), & |g_i(t)| \leqslant \alpha_i(t), \\ \alpha_i(t) \operatorname{sign} g_i(t), & |g_i(t)| > \alpha_i(t) \end{cases}$$
$$(i = 1, \ldots, r).$$

2) For U_2,

$$v(t) = \begin{cases} G(t), & G^*(t) N(t) G(t) \leqslant \beta(t), \\ [E + \lambda(t) N(t)]^{-1} G(t), & G^*(t) N(t) G(t) > \beta(t), \end{cases} \tag{3.12}$$

where $\lambda(t)$ satisfies the equation

$$G^*(t) \left[(E + \lambda(t) N(t))^{-1} N(t) (E + \lambda(t) N(t))^{-1} \right] G(t) = \beta(t).$$

3) For U_3, equation (3.12) is valid with $N(t) \equiv N$ and $\beta(t) \equiv 1$.
4) For U_4,

$$v(t) = \begin{cases} G(t), & \|G(t)\| \leqslant 1, \\ \dfrac{G(t)}{\|G(t)\|}, & \|G(t)\| > 1. \end{cases}$$

5) For U_5,

$$v^i(t) = \begin{cases} g_i(t), & |g_i(t)| \leqslant 1, \\ \operatorname{sign} g_i(t), & |g_i(t)| > 1 \end{cases}$$
$$(i = 1, \ldots, r).$$

6) For U_6,

$$v^i(t) = a_{ij_i} \quad (i = 1, \ldots, r),$$

where

$$(a_{ij_i} - g_i(t))^2 = \min_{j=1, \ldots, p_i} (a_{ij} - g_i)^2.$$

7) For U_7,

$$v^i(t) = \begin{cases} g_i(t), & \|g_i\|^2 \leqslant C_i, \\ \dfrac{\sqrt{C} g_i(t)}{\|g_i\|}, & \|g_i\|^2 > C_i \end{cases} \quad (i = 1, \ldots, r).$$

8) For U_8,

$$v(t) = \begin{cases} G(t), & \|G\|_N^2 \leqslant C, \\ (E + \lambda N(t))^{-1} A(t), & \|G\|_N^2 > C, \end{cases}$$

where

$$\|G\|_N^2 = \int_0^T G^*(t) N(t) G(t)\, dt,$$

and λ satisfies the equation

$$\int_0^T G^*(t) [(E + \lambda N(t))^{-1} N(t) (E + \lambda N(t))^{-1}] G(t)\, dt = C.$$

9) For U_9,

$$v(t) = \begin{cases} G(t), & \|G\|^2 \leqslant 1, \\ \dfrac{G(t)}{\|G\|}, & \|G\|^2 > 1. \end{cases}$$

10) For U_{10}, if we reason as in subsection 2°, we can easily reduce our problem to that of minimizing, for some $\lambda > -1$, the functional

$$\int_0^T [(u(t) - a(t))^2 + \lambda u^2(t)]\, dt - (1 + \lambda) C =$$
$$= \int_0^T [-2a(t) u(t) + (1 + \lambda) u^2(t) + a^2(t)]\, dt - (1 + \lambda) C \quad (3.13)$$

on the set

$$U' = \{u \in L^2([0, T]) \mid |u(t)| \leqslant 1\}.$$

(In (3.13), $a(t)$ is a measurable bounded function on $[0, T]$.) The minimum of the functional (3.13) is attained with

$$u_\lambda(t) = \operatorname{sat} \frac{a(t)}{1+\lambda},$$

where the function sat x is defined by formula (3.5). The problem reduces to solving an equation just as in 2°:

$$\int_0^T \left(\operatorname{sat} \frac{a(t)}{1+\lambda} \right)^2 dt = C.$$

CHAPTER 5

Some Problems in Finite-Dimensional Spaces

1. The second-order methods.

In Chapter 3, we considered some methods of finding stationary points of functionals that employ only the gradients of functions or first derivatives with respect to direction.

Below, we shall show for the case of Euclidean spaces that the methods expounded above can be considered to be discrete variants of "continuous" methods and we shall also show how one can construct procedures that employ second derivatives of the function to be minimized (the so-called, second-order methods).

We shall not be concerned with methods of random search or methods that employ for the construction of X_{k+1} not only X_k but also X_{k-1}, X_{k-2}, etc.

We note that the finite dimensionality of the space in question is essential since we shall need to consider differential equations the right-hand number of which, while continuous, does not in general satisfy a Lipschitz condition. That such equations have a solution is ensured by Peano's theorem, which is valid only in the finite-dimensional case (see [4, Chapter 1, Sec. 1, 3°]).

In the present chapter, we shall denote n-dimensional Euclidean space by E_n.

1°. Let $f(X)$ denote a continuously differentiable function defined on E_n. For $f(X)$ to attain its minimum value on E_n at a point $\widetilde{X} \in E_n$, it is necessary (and if $f(X)$ is convex, sufficient) that

$$\frac{\partial f(\widetilde{X})}{\partial X}=0, \tag{1.1}$$

where $\frac{\partial f}{\partial X}$ denotes the gradient of f. If (1.1) holds at some point $X \in E_n$, then X is called a *stationary point*.

Consider the system of differential equations

$$\frac{dX(t)}{dt}=-\frac{\partial f(X(t))}{\partial X}, \tag{1.2}$$

$$X(0)=X_0. \tag{1.3}$$

We shall denote by $X(t, X_0)$ the solution of the system (1.2) with initial condition (1.3). In accordance with Peano's theorem, a solution exists and is continuous. If the function $\frac{\partial f(X)}{\partial X}$ satisfies a Lipschitz condition, this solution is unique. If the set

$$M(X_0)=\{X \mid X \in E_n,\ f(X) \leqslant f(X_0)\} \tag{1.4}$$

is bounded, there exists a convergent sequence of the points X_k:

$$X_k=X(t_k, X_0), \quad t_k \xrightarrow[k\to\infty]{} \infty, \quad X_k \to X^* \in M(X_0).$$

(Since $M(X_0)$ is bounded, an arbitrary sequence $\{X(t_k)\}$ such that $t_k \to \infty$ contains a convergent subsequence.) Then, X^* is a stationary point and, if $f(X)$ is a convex function, then X^* is a minimum of the function $f(X)$ on E_n.

Thus, for an arbitrary initial condition X_0 for which the set $M(X_0)$ defined by (1.4) is bounded, a solution of the system (1.2) approaches a stationary point. On the basis of this "continuous" method of minimization of the function $f(X)$, we can construct discrete methods. In particular, we can assume that the gradient method is a numerical method of solving the system of differential equations (1.2).

On the other hand, suppose that $f(X)$ is twice continuously differentiable. It follows from the necessary condition (1.1) that the function

$$F(X)=\frac{1}{2}\left(\frac{\partial f(X)}{\partial X}\right)^2 \tag{1.5}$$

also attains its minimim value (namely, 0) at the point $\widetilde{X}$. Obviously, all stationary points of the function $f(X)$ are minima of the function $F(X)$ and all minima of the function $F(X)$ are stationary points of the function $f(X)$. If $f(X)$ is a convex function, then a minimum of $F(X)$ is a minimum of the function $f(X)$.

Consider the system of differential equations

$$\frac{dX(t)}{dt} = -\frac{\partial^2 f(X(t))}{\partial X^2}\frac{\partial f(X(t))}{\partial X}, \tag{1.6}$$

$$X(0) = X_0. \tag{1.7}$$

If the set $M_1(X_0) = \{X \mid X \in E_n,\ F(X) \leqslant F(X_0)\}$ is bounded, then an arbitrary solution of the system (1.6), (1.7) approaches a stationary point of the function $F(X)$. Therefore, to find the stationary points of the function $f(X)$, we can construct algorithms by using the matrix of the second derivatives (the so-called, second-order methods).

The direction of fastest decrease of the function $f(X)$ at the point X is the direction $X_1(X) = -\frac{\partial f}{\partial X}$, and the direction of fastest decrease of $F(X)$ is the direction

$$X_2(X) = -\frac{\partial^2 f(X)}{\partial X^2}\frac{\partial f(X)}{\partial X}.$$

These two directions make it possible to construct second-order algorithms. We present one such method. Suppose that we have found X_k. Let us set

$$X_{1k} = X_1(X_k), \quad X_{2k} = X_2(X_k), \quad X_k(\alpha,\ \beta) = X_k + \alpha X_{1k} + \beta X_{2k}.$$

Let us find $\alpha_k \in [0,\ \infty)$ and $\beta_k \in [0,\ \infty)$ such that

$$f(X_k(\alpha_k,\ \beta_k)) = \min_{\substack{\alpha \in [0,\ \infty) \\ \beta \in [0,\ \infty)}} f(X_k(\alpha,\ \beta)).$$

Let us set

$$X_{k+1} = X_k(\alpha_k,\ \beta_k).$$

Obviously, $f(X_{k+1}) \leqslant f(X_k)$; also, if $X_{1k} \neq 0$ (that is, if the point X_k is not stationary), then $f(X_{k+1}) < f(X_k)$. The sequence of approximations thus constructed converges to a stationary point of the function $f(X)$. However, at every stage, we need to minimize the function of two variables

$$f_k(\alpha, \beta) = f(X_k(\alpha, \beta)).$$

It is possible to construct second-order methods that do not require minimization of a function of two variables at every stage.

Solution of the system (1.2), (1.3) is a curve $X(t, X_0)$ with parameter t. The method of fastest descent and the gradient method are based on replacement of this curve with its tangent. If $f(X)$ is a twice continuously differentiable function, we can approximate this curve more precisely. We have

$$X(t, X_0) = X(0, X_0) + tX_1(0, X_0) + \frac{1}{2} t^2 X_2(0, X_0) + o(t^2),$$

where $o(t^2)$ is a vector such that

$$\frac{\| o(t^2) \|}{t^2} \xrightarrow[t \to 0]{} 0,$$

$$X_1(0, X_0) = \left. \frac{dX(t, X_0)}{dt} \right|_{t=0} = -\frac{\partial f(X_0)}{\partial X},$$

$$X_2(0, X_0) = \left. \frac{d^2 X(t, X_0)}{dt^2} \right|_{t=0} = -\frac{\partial^2 f(X_0)}{\partial X^2} \frac{\partial f(X_0)}{\partial X}.$$

From this we immediately obtain a method of successive approximations.

Let X_1 denote an arbitrary point in E_n and suppose that we have found X_k. Define $X_{1k} = X_1(0, X_k)$, $X_{2k} = X_2(0, X_k)$;

$$X_k(t) = X_k + tX_{1k} + \frac{1}{2} t^2 X_{2k}.$$

Let us find t_k such that

$$f(X_k(t_k)) = \min_{t \in [0, \infty)} f(X_k(t)), \tag{1.8}$$

and let us set

$$X_{k+1} = X_k(t_k). \tag{1.9}$$

Obviously, $f(X_{k+1}) \leqslant f(X_k)$. Also, if $X_{1k} \neq 0$ (that is, if $\frac{\partial f(X_k)}{\partial X} \neq 0$), then $f(X_{k+1}) < f(X_k)$. The sequence $\{X_k\}$ thus constructed converges to a stationary point of the function $f(X)$ on E_n.

Just as in the gradient method, we can, under certain additional restrictions on $f(X)$, find ε_1 and ε_2 such that $\varepsilon_2 > \varepsilon_1 > 0$ and for X_{k+1} we can take $X_k(t_k)$, where t_k is any number in the interval $[\varepsilon_1, \varepsilon_2]$.

Remark: There is another well-known method, one that uses both the first and the second derivatives of the function $f(X)$ to find the next approximation. This is Newton's method, which is based on replacing the function $f(X)$ with a second-order approximation of it. Instead of the function $f(X)$, at the kth stage one minimizes the function

$$f_k(X) = f(X_k) + \left(X, \frac{\partial f(X_k)}{\partial X}\right) + \frac{1}{2}\left(X, \frac{\partial^2 f(X_k)}{\partial X^2} X\right),$$

and takes for X_{k+1} the point staisfying the condition

$$f_k(X_{k+1}) = \min_{X \in E_n} f_k(X_k).$$

In [39], this method is extended to the case in which the minimum $f(X)$ is sought on a bounded set. Under rather stringent restrictions, Newton's method converges. Apart from quite serious difficulties associated with minimizing the function $f_k(X)$ [especially on a bounded set], the method is inconvenient in that $f(X_{k+1})$ may prove to be greater than $f(X_k)$. Furthermore, the interval $[X_k, X_{k+1}]$ does not include a point X such that $f(X) < f(X_k)$ although the point X_k is not a stationary point of the function $f(X)$.

The method (1.8) is convenient in that, first of all, at each stage it is necessary to minimize a function of a single variable and, second, the values of the function decrease monotonically and the method ensures (for all twice continuously differentiable functions) convergence to a stationary point.

2°. Let $f(X)$ denote a continuously differentiable function defined on a closed convex bounded subset Ω of E_n. Then, for the function $f(X)$ to attain its minimum value on Ω at a point $\widetilde{X} \in \Omega$, it is necessary (and, if $f(X)$ is convex, sufficient) that (see II.2.11)

$$\varphi(\widetilde{X}) = \min_{X \in \Omega}\left(\frac{\partial f(\widetilde{X})}{\partial X}, X - \widetilde{X}\right) = 0. \tag{1.10}$$

A point $\widetilde{X}$ satisfying (1.10) is called a stationary point of the function $f(X)$ on the set Ω. For a convex function, an arbitrary stationary point is a minimum. Condition (1.10) means that, if $\widetilde{X}$ is an interior point of Ω, then

$$\frac{\partial}{\partial X} f(\widetilde{X}) = 0,$$

On the other hand, if $\widetilde{X}$ is a boundary point of Ω, then either

$$\frac{\partial}{\partial X} f(\widetilde{X}) = 0,$$

or the hyperplane

$$\left(\frac{\partial f(\widetilde{X})}{\partial X}, X - \widetilde{X}\right) = 0$$

is the support hyperplane for the set Ω at the point $\widetilde{X}$.

Consider the function

$$\varphi(X) = \min_{Z \in \Omega} \left(\frac{\partial f(X)}{\partial X}, Z - X\right). \qquad (1.11)$$

Obviously, $\varphi(X) \leqslant 0$. Therefore, an arbitrary stationary point of the function $f(X)$ on the set Ω is a maximum of the function $\varphi(X)$, and an arbitrary point representing a maximum of $\varphi(X)$ on the set Ω is a stationary point of the function $f(X)$ on the set Ω. If $f(X)$ is a twice continuously differentiable function, then $\varphi(X)$ is differentiable with respect to direction and

$$\varphi'_X(g) = \lim_{\alpha \to +0} \frac{\varphi(X + \alpha g) - \varphi(X)}{\alpha} =$$
$$= \min_{Z \in R(X)} \left[\frac{\partial^2 f(X)}{\partial X^2}(Z - X) - \frac{\partial f(X)}{\partial X}, g\right], \qquad (1.12)$$

where $g \in E_n$, $g \neq 0$, and $Z \in R(X)$, if

$$\varphi(X) = \left(\frac{\partial f(X)}{\partial X}, Z - X\right).$$

We note that, if Ω is a strictly convex set and if the point X is not a stationary point of the function $f(X)$ on Ω, then $R(X)$ consists of a single point, which we shall also denote by $R(X)$. In this case, the function $\varphi(X)$ is continuously differentiable at an arbitrary non-stationary point of X and

$$\frac{\partial \varphi(X)}{\partial X} = \frac{\partial^2 f(X)}{\partial X^2}(Z - X) - \frac{\partial f(X)}{\partial X},$$

where $Z = R(X)$.

Thus, the problem of finding the stationary points of the function $f(X)$ on the set Ω is equivalent to finding points representing a maximum of the function $\varphi(X)$.

For a function $\varphi(X)$ to attain its maximum value on Ω at a point $X' \in \Omega$, it is necessary and, in case $f(X)$ is convex, sufficient that (see II.2.11)

$$\psi(X) = \max_{V \in \Omega}\left[\min_{Z \in R(X)}\left(\frac{\partial^2 f(X)}{\partial X^2}(Z - X) - \frac{\partial f(X)}{\partial X},\ V - X\right)\right] \equiv$$
$$\equiv \max_{V \in \Omega}\ \min_{Z \in R(X)} \Phi(Z,\ V,\ X) = 0. \tag{1.13}$$

It can be shown that, if $f(X)$ is a convex function, then the set of points at which $f(X)$ attains a minimum on the set Ω coincides with the set of points at which the function $\varphi(X)$ has a maximum on Ω.

We note that, if φ is a convex function, then, for $X \in \Omega$ and $Z \in R(X)$,

$$\psi(X) \geqslant \Phi(Z,\ Z,\ X) = \left(Z - X,\ \frac{\partial^2 f(X)}{\partial X^2}(Z - X)\right) -$$
$$- \left(\frac{\partial f(X)}{\partial X},\ Z - X\right) \geqslant 0,$$

because the first term is nonnegative by virtue of the convexity of $f(X)$ and

$$\left(\frac{\partial f(X)}{\partial X},\ Z - X\right) = \varphi(X) \leqslant 0.$$

Let X denote a member of Ω and suppose that $\frac{\partial f(X)}{\partial X} \neq 0$. The point on the straight line $X - \alpha G(X)$ closest to a point $Y \in E_n$ is the point

$$Z(Y) = X - \beta G(X),$$

where

$$\beta = -\frac{2(Y - X, G(X))}{\|G(X)\|^2}, \quad G(X) = \frac{\partial f(X)}{\partial X}.$$

Let θ denote a member of $R(X)$ such that

$$(\theta - Z(\theta))^2 = \min_{Y \in R(X)} (Y - Z(Y))^2. \tag{1.14}$$

The point $\theta = \theta(X)$ is unique. If $f(X)$ is a continuously differentiable function and Ω is a strictly convex set, then $\theta(X)$ is a continuous vector-valued function.

Let us look at the system of ordinary differential equations

$$\frac{dX(t)}{dt} = \theta(X(t)) - X(t), \tag{1.15}$$

$$X(0) = X_0. \tag{1.16}$$

A system of a more complicated form than (1.15) was studied in detail in Sec. 2. The right-hand member of (1.15) is continuous if the set Ω is strictly convex. Therefore, in accordance with Peano's theorem, the system (1.15) always has a solution $X(t, X_0)$ though it may not be unique if the right-hand member does not satisfy a Lipschitz condition. Also, $X(t, X_0)$ is a vector-valued function that is continuous with respect to t on $[0, \infty)$. If $X_0 \in \Omega$, then $X(t, X_0) \in \Omega$ for all $t \in [0, \infty)$.

If the set Ω is strongly convex and if X_0 is a member of Ω, then the limit of every convergent sequence $\{X_k\}$, where $X_k = X(t_k, X_0)$ and $t_k \to \infty$, is a stationary point of the function $f(X)$ on the set Ω. From this it is obvious that all schemes of the conditional-gradient method are methods of solving the system (1.15). The projective methods studied in Chapter 3 can be obtained by considering systems of differential equations similar to (1.15). (In this case, the point $\theta(X(t))$ must be obtained in a somewhat different manner.)

Thus, we see that methods for finding stationary points of continuously differentiable functions on a set Ω that are based on use of first derivatives of the function $f(X)$ only can be obtained by considering the system of differential equations (1.15).

By using the function $\varphi(X)$ defined by (1.11), we can construct second-order methods. The function $\varphi(X)$ is differentiable with respect to direction and, to find its maximum on Ω (a maximum of this function is a stationary point of the function $f(X)$ on Ω), it is necessary to use a method analogous to that expounded in subsection 6° of Sec. 1 of Chapter 3. If ii is a strictly convex set, then $\varphi(X)$ is continuously differentiable. Second-order methods using simultaneously the direction of decrease of the function $f(X)$ and the direction of increase of the function $\varphi(X)$ are of interest.

Suppose that we have found X_k. For the point X_k, let us find the points

$$\theta_k = \theta(X_k) \in \Omega$$

and $V_k \in \Omega$, where

$$\min_{Z \in R(X)} \Phi(Z, V_k, X_k) = \max_{V \in \Omega} \min_{Z \in R(X_k)} \Phi(Z, V, X_k).$$

Then, the point

$$X_k(\alpha, \beta) = (1 - \alpha - \beta) X_k + \alpha\theta_k + \beta V_k,$$

where $\alpha \in [0, 1]$, $\beta \in [0, 1]$, and $\alpha + \beta \leqslant 1$.

Let us find $\alpha_k \geqslant 0$ and $\beta_k \geqslant 0$ such that $\alpha_k + \beta_k \leqslant 1$ and

$$f(X_k(\alpha_k, \beta_k)) = \min_{\substack{\alpha, \beta \geqslant 0 \\ \alpha + \beta \leqslant 1}} f(X_k(\alpha, \beta)),$$

and let us set

$$X_{k+1} = X_k(\alpha_k, \beta_k). \tag{1.17}$$

Here, $f(X_{k+1}) \leqslant f(X_k)$; also, if $\varphi(X_k) < 0$, then $f(X_{k+1}) < f(X_k)$. The sequence $\{X_k\}$ constructed according to (1.17) converges to a stationary point of the function $f(X)$ on Ω. Here, we need to minimize the function

$$f_k(\alpha, \beta) \equiv f(X_k(\alpha, \beta))$$

on the square

$$[0, 1] \times [0, 1].$$

It is possible to construct yet other second-order methods that require at each stage minimization of a function of only one variable (for example, we may take $\alpha = \beta$).

3°. Finally, let us look at the problem of finding

$$\min_{Y \in E_m} \max_{X \in E_n} f(X, Y),$$

where the function $f(X, Y)$ is continuous with respect to X and continuously differentiable with respect to Y on $E_n \times E_m$, where

$$\left\| \frac{\partial f(X, Y)}{\partial Y} \right\|$$

is bounded on $E_n \times E_m$, where the set $M_X = \{X \mid f(X, Y) = \varphi(Y)\}$ is bounded for some $Y \in E_m$, and where, for some $Y \in E_m$, there is no sequence of points $\{X_i\}$ such that $X_i \in E_n, \|X_i\| \to \infty$, and

$$f(X_i, Y) \to \sup_{X \in E_n} f(X, Y).$$

From formulas (II.2.19) and (II.3.33), we can easily show the validity of the following assertion [cf. (II.2.29)]: For the function

$$\varphi(Y) = \max_{X \in E_n} f(X, Y)$$

to attain its maximum value on E_n at a point $Y \in X_m$, it is necessary (and, if $\varphi(Y)$ is convex, sufficient) that

$$\psi_1(Y) = \min_{\substack{\|g\|=1 \\ g \in E_m}} \frac{\partial \varphi(Y)}{\partial g} = \min_{\substack{\|g\|=1 \\ g \in E_m}} \max_{X \in R(Y)} \left(\frac{\partial f(X, Y)}{\partial Y}, g \right) \geqslant 0, \qquad (1.18)$$

where $X \in R(Y)$ if

$$f(X, Y) = \varphi(Y) = \max_{Z \in E_n} f(Z, Y).$$

The condition (1.18) is equivalent to the condition

$$\psi_2(Y) = \min_{\substack{\|g\| \leqslant 1 \\ g \in E_m}} \max_{X \in R(Y)} \left(\frac{\partial f(X, Y)}{\partial Y}, g \right) = 0. \tag{1.19}$$

A point $Y \in E_m$ that satisfies (1.18) is called a stationary point of the function $\varphi(Y)$ on E_m.

Condition (1.18) means geometrically that the convex hull $L(Y)$ of the set of vectors

$$H(Y) = \left\{ \frac{\partial f(X, Y)}{\partial Y} \,\middle|\, X \in R(Y) \right\},$$

must, at a point minimizing the function $\varphi(Y)$, include the coordinate origin (cf. subsection 16°, Sec. 2, Chapter 2).

If a point $Y \in E_m$ is not a stationary point, then $\psi_1(Y) = \psi_2(Y)$. On the other hand, if $Y \in E_m$ is a stationary point, then $\psi_2(Y) = 0$, and $\psi_2(Y) = r(Y)$, where $r(Y)$ is the radius of the largest sphere with center at the coordinate origin that is contained in the interior of the set $L(Y)$.

Suppose that a function $f(X, Y)$ is, for arbitrary fixed X, convex with respect to Y. Let us choose some $Y \in E_m$. Suppose that a set $\omega \subset E_n$ is such that

$$\min_{\|g\| \leqslant 1} \max_{X \in \omega} \left(\frac{\partial f(X, Y)}{\partial Y}, g \right) = 0.$$

Then,

$$\min_{X \in \omega} f(X, Y) \leqslant \min_{Z \in E_m} \varphi(Z) \leqslant \varphi(Y). \tag{1.20}$$

To see this, note that, obviously,

$$\min_{Z \in E_n} \varphi(Z) \leqslant \varphi(Y).$$

Let us prove the first of inequalities (1.20). If there existed a point Y such that

$$\varphi(Y') < \min_{X \in \omega} f(X, Y),$$

we would have

$$f(X, Y') \leqslant \varphi(Y') = \max_{V \in E_n} f(V, Y'),$$

and, for $X \in \omega$, we would have $f(X, Y') < f(X, Y) - a$, where $a > 0$. Since $f(X, Y)$ is convex with respect to Y, we have

$$\left(\frac{\partial f(X, Y)}{\partial Y}, Y' - Y\right) \leqslant -\rho < 0$$

for $X \in \omega$. Then,

$$\max_{X \in \omega} \left(\frac{\partial f(X, Y)}{\partial Y}, g'\right) < 0,$$

where

$$g' = \frac{Y' - Y}{\| Y' - Y \|},$$

which contradicts the assumption. This proves (1.20).

What has been said gives us a convenient method for approximate location of the points at which the function $\varphi(Y)$ has a minimum. (Inequality (1.20) provides a bound on the distance to an extremum.)

We note that the function $\psi_2(Y) \leqslant 0$ and that an arbitrary stationary point of $\varphi(Y)$ is a point at which $\psi_2(Y)$ has a maximum. It is also obvious that an arbitrary point at which the function $\psi_2(Y)$ has a maximum is at the same time a stationary point of the function $\varphi(Y)$.

In subsection 6° of Sec. 1 of Chapter 3, we considered a method of successive approximations that, in particular, was suitable for finding stationary points of the function $\varphi(Y)$. In using this method to solve our problem, we need, at the kth stage, to choose as direction of descent a direction g_k such that

$$\psi_{2k}(Y_k) = \max_{X \in R_{\varepsilon_k}(Y_k)} \left(\frac{\partial f(X, Y_k)}{\partial Y}, g_k\right) =$$

$$= \min_{\substack{\|g\|=1 \\ g \in E_m}} \max_{X \in R_{\varepsilon_k}(Y_k)} \left(\frac{\partial f(X, Y_k)}{\partial Y}, g\right), \tag{1.21}$$

where ε_k is a positive number independent of k and $X \in R_{\varepsilon_k}(Y_k)$ whenever

$$\varphi(Y_k) - f(X, Y_k) \leqslant \varepsilon_k.$$

If Y_k is not a stationary point, then

$$\psi_{2\varepsilon_k}(Y_k) < 0.$$

In this case, the direction g_k satisfying (1.21) is unique and we have $\|g\| = 1$.

It is possible to find another direction of descent. For this direction, we take the direction of "steepest" ascent for the function

$$\psi_{2k}(Y) = \min_{\substack{\|g\|=1 \\ g \in E_m}} \max_{X \in R_{\varepsilon_k}(Y)} \left(\frac{\partial f(X, Y)}{\partial Y}, g\right)$$

at the point Y_k.

The question of the differentiability of the function $\psi_{2k}(Y)$ will be considered in Sec. 4. By applying formulas (4.12) and (4.11) and remembering that $\|g\| \leqslant 1$, we have

$$(\psi_{2k})'_{Y_k}(Z) \equiv \frac{\partial \psi_{2k}(Y_k)}{\partial Z} =$$

$$= \min_{V \in A(g_k)} \max_{X \in B(g_k, Y_k)} \left[\left(\frac{\partial f(X, Y_k)}{\partial Y}, V\right) + \left(\frac{\partial^2 f(X, Y_k)}{\partial Y^2} g_k, Z\right)\right], \quad (1.22)$$

$V \in A(g_k)$ if $(V, g_k) \leqslant 0$, $V \in E_m$; $X \in B(g_k, Y_k)$ if $X \in R_{\varepsilon_k}(Y_k)$ and

$$\left(\frac{\partial f(X, Y_k)}{\partial Y}, g_k\right) = \max_{Z \in R_{\varepsilon_k}(Y_k)} \left(\frac{\partial f(X, Y_k)}{\partial Y}, g_k\right). \quad (1.23)$$

As the second direction, we can choose Z_k such that $\|Z_k\| = 1$:

$$(\psi_{2k})'_{Y_k}(Z_k) \equiv \frac{\partial \psi_{2k}(Y_k)}{\partial Z_k} = \max_{\|Z\| \leqslant 1} \frac{\partial \psi_{2k}(Y_k)}{\partial Z}. \quad (1.24)$$

Remark. As the auxiliary direction, we can also take the direction Z_k satisfying the equation

$$\frac{\partial \psi_2(Y_k)}{\partial Z_k} = \max_{\|Z\| \leqslant 1} \frac{\partial \psi_2(Y_k)}{\partial Z}. \quad (1.25)$$

2. The finding of saddle points.

In the present section, we shall consider the problem of finding saddle points of a continuously differentiable function $f(X, Y)$. Another approach to the solution of these problems is to be found in [1, 8, 12, 33, 50, 58]. In subsection 2°, we shall consider the case in which the ranges of values of the variables of X and Y are respectively n- and m-dimensional Euclidean spaces. In subsection 3°, we shall study the case in which the ranges of values of the variables X and Y are bounded sets.

1°. Let $f(X, Y)$ denote a function that is continuously differentiable on $\Omega_X \times \Omega_Y$. Let Ω_X denote a subset of E_n and let Ω_Y denote a subset of E_m. Suppose that these two subsets are closed and bounded.

A point $[X^*, Y^*]$ is called a *saddle point* of a function $f(X, Y)$ on the set $\Omega_X \times \Omega_Y$ if, for all $X \in \Omega_X$ and $Y \in \Omega_Y$,

$$f(X, Y^*) \leqslant f(X^*, Y^*) \leqslant f(X^*, Y). \qquad (2.1)$$

Then,

$$f(X^*, Y^*) = \max_{X \in \Omega_X} f(X, Y^*) = \min_{Y \in \Omega_Y} f(X^*, Y). \qquad (2.2)$$

We shall say that a function $f(X, Y)$ is concave-convex if, for arbitrary fixed $Y \in \Omega_Y$, the function $f_Y(X) \equiv f(X, Y)$ is concave with respect to X on Ω_X and, for arbitrary fixed $X \in \Omega_X$, the function $f_X(Y) \equiv f(X, Y)$ is convex with respect to Y on Ω_Y.

In a natural manner, we can define a strictly concave-convex function. We wish to find a saddle point of the function $f(X, Y)$ on $\Omega_X \times \Omega_Y$.

From the necessary condition (II.2.11), we get

THEOREM 2.1. *For a point $[X^*, Y^*]$ to be a saddle point of the function $f(X, Y)$ on the set $\Omega_X \times \Omega_Y$, it is necessary (and if $f(X, Y)$ is concave-convex on $\Omega_X \times \Omega_Y$, sufficient) that*

$$\max_{X \in \Omega_X} \left(\frac{\partial f(X^*, Y^*)}{\partial X}, (X - X^*) \right) = \min_{Y \in \Omega_Y} \left(\frac{\partial f(X^*, Y^*)}{\partial Y}, (Y - Y^*) \right) = 0. \qquad (2.3)$$

COROLLARY. *If $\Omega_X = E_n$ and $\Omega_Y = E_n$ condition (2.3) can be replaced with the condition*

$$\frac{\partial f(X^*, Y^*)}{\partial X} = \frac{\partial f(X^*, Y^*)}{\partial Y} = 0. \tag{2.4}$$

A point $[X^*, Y^*] \in \Omega_X \times \Omega_Y$ that satisfies (2.3) or [if the conditions of the corollary are satisfied] (2.4), is called a stationary point of the function $f(X, Y)$ on the set $\Omega_X \times \Omega_Y$.

2°. Suppose that $\Omega_X = E_n$ and $\Omega_Y = E_m$. Consider the systems of differential equations:

$$\frac{dX(t)}{dt} \equiv \dot{X}(t) = \frac{\partial f(X, Y)}{\partial X}, \tag{2.5}$$

$$X(0) = X_0 \in E_n, \tag{2.6}$$

$$\frac{dY(t)}{dt} \equiv \dot{Y}(t) = -\frac{\partial f(X, Y)}{\partial Y}, \tag{2.7}$$

$$Y(0) = Y_0 \in E_m. \tag{2.8}$$

By virtue of the continuity of the functions in the right-hand members of these four equations, it follows from Peano's theorem that the systems (2.5) and (2.7) have at least one solution.

Let us denote by $X(t, X_0, Y_0)$ and $Y(t, X_0, Y_0)$ the solutions of the systems (2.5) and (2.7) with the initial conditions (2.6) and (2.8). Let us suppose that the function $f(X, Y)$ is twice continuously differentiable and strongly convex-concave on $E_n \times E_m$. Then, matrices of $-\frac{\partial^2 f}{\partial X^2}$ and $\frac{\partial^2 f}{\partial Y^2}$ are strictly positive-definite; that is, for arbitrary

$$(Z_1, Z_2) \in E_n \times E_m \text{ and } (X, Y) \in E_n \times E_m$$

we have

$$-Z_1^* \left(\frac{\partial^2 f(X, Y)}{\partial X^2} Z_1 \right) \geqslant m_1(X, Y) \| Z_1 \|^2, \quad m_1(X, Y) > 0, \tag{2.9}$$

$$Z_2^* \left(\frac{\partial^2 f(X, Y)}{\partial Y^2} Z_2 \right) \geqslant m_2(X, Y) \| Z_2 \|^2, \quad m_2(X, Y) > 0. \tag{2.10}$$

Also, for an arbitrary bounded set $S \subset E_n \times E_m$, there exist positive numbers m_1 and m_2 (dependent on S) such that $m_1(X, Y) \geqslant m_1 > 0$ and $m_2(X, Y) \geqslant m_2 > 0$ for all $[X, Y] \in S$. We denote by

$$M(X_0, Y_0) \subset E_n \times E_m$$

the set

$$\{(X, Y) \mid F(X, Y) \leqslant F(X_0, Y_0)\},$$

where

$$F(X, Y) = \frac{1}{2}\left[\left(\frac{\partial f(X, Y)}{\partial X}\right)^2 + \left(\frac{\partial f(X, Y)}{\partial Y}\right)^2\right].$$

Under these assumptions, we have

THEOREM 2.2. *If the set $M(X_0, Y_0)$ is bounded, then the solutions $X(t, X_0, Y_0)$ and $Y(t, X_0, Y_0)$ of the systems (2.5) and (2.7) converge to a unique saddle point.*

Proof: Let us show first that

$$F(t) \equiv F(X(t, X_0, Y_0), Y(t, X_0, Y_0)) \underset{t \to \infty}{\rightarrow} 0. \tag{2.11}$$

Since

$$F(t) = F(X_0, Y_0) + \int_0^t \left[\left(\frac{\partial f(\tau)}{\partial X}, \frac{\partial^2 f(\tau)}{\partial X^2} \frac{dX(\tau)}{d\tau}\right) + \right.$$
$$+ \left(\frac{\partial f(\tau)}{\partial X}, \frac{\partial^2 f(\tau)}{\partial X \partial Y} \frac{dY(\tau)}{d\tau}\right) + \left(\frac{\partial f(\tau)}{\partial Y}, \frac{\partial^2 f(\tau)}{\partial Y \partial X} \frac{dX(\tau)}{d\tau}\right) +$$
$$\left. + \left(\frac{\partial f(\tau)}{\partial Y}, \frac{\partial^2 f(\tau)}{\partial Y^2} \frac{dY(\tau)}{d\tau}\right)\right] d\tau, \tag{2.12}$$

where

$$f(\tau) \equiv f(X(\tau, X_0, Y_0), Y(\tau, X_0, Y_0)),$$

when we substitute the values $\frac{dX(\tau)}{d\tau}$ and $\frac{dY(\tau)}{d\tau}$ found from (2.5) and (2.7) into (2.12), we obtain

$$F(t) = F(X_0, Y_0) + \int_0^t \left[\left(\frac{\partial f(\tau)}{\partial X}, \frac{\partial^2 f(\tau)}{\partial X^2} \frac{\partial f(\tau)}{\partial X}\right) - \right.$$
$$\left. - \left(\frac{\partial f(\tau)}{\partial Y}, \frac{\partial^2 f(\tau)}{\partial Y^2} \frac{\partial f(\tau)}{\partial Y}\right)\right] d\tau. \tag{2.13}$$

By virtue of (2.9) and (2.10), we have $F(t) \leqslant F(X_0, Y_0)$. Therefore for arbitrary $t \geqslant 0$, the point $[X(t, X_0, Y_0), Y(t, X_0, Y_0)]$ belongs to $M(X_0, Y_0)$. By virtue of the boundedness of $M(X_0, Y_0)$, the solutions of the systems (2.5) and (2.7) are also bounded.

Now let us suppose that (2.11) does not hold. Then, there exists a sequence $\{t_k\}$ such that

$$t_k \xrightarrow[k\to\infty]{} \infty$$

and

$$F(t_k) \equiv F(X(t_k, X_0, Y_0), Y(t_k, X_0, Y_0)) \geqslant \rho > 0.$$

Suppose, for example, that there exist infinitely many k such that

$$\left\| \frac{\partial f(X_k, Y_k)}{\partial X} \right\| \geqslant \rho' > 0,$$

where

$$X_k = X(t_k, X_0, Y_0), \quad Y_k = Y(t_k, X_0, Y_0).$$

Then, by virtue of the properties of the right-hand members of the systems (2.5) and (2.7), for arbitrary $\varepsilon > 0$ there exists a $\delta > 0$ such that

$$\|X(t') - X(t)\| \leqslant \varepsilon, \quad \|Y(t') - Y(t)\| \leqslant \varepsilon,$$

whenever $|t' - t| \leqslant \delta$.

Therefore, there exists a $\mu > 0$ such that

$$\left\| \frac{\partial f(X(t), Y(t))}{\partial X} \right\| \geqslant \frac{1}{2}\rho' \tag{2.14}$$

for $|t - t_k| \leqslant \mu$.

Let us take the subsequence of $\{t_k\}$ that includes only those elements for which $t_{k+1} - t_k \geqslant 2\mu$, $k = 1, 2, \ldots$.

But then,

$$F(X(t), Y(t)) \leqslant F(X_0, Y_0) - m_1 \sum_{k=1}^{s(t)} \int_{t_k-\mu}^{t_k+\mu} \left\| \frac{\partial f(X(\tau), Y(\tau))}{\partial X} \right\|^2 d\tau, \tag{2.15}$$

where $s(t)$ is a natural number such that

$$t_{s(t)}+\mu\leqslant t\leqslant t_{s(t)+1}+\mu,$$

and $m_1>0$ corresponds to the set $M(X_0, Y_0)$. Keeping (2.14) in mind, we obtain from (2.15)

$$F(X(t), Y(t))\leqslant F(X_0, Y_0)-\frac{1}{2}m_1, \mu\rho'^2 s(t).$$

Since $s(t)\underset{t\to\infty}{\longrightarrow}+\infty$ and $\frac{1}{2}m_1\mu\rho'^2>0$ is independent of t, we have

$$F(X(t), Y(t))\underset{t\to\infty}{\longrightarrow}-\infty,$$

which is impossible since the continuous function $F(X, Y)$ is bounded on the bounded set $M(X_0, Y_0)$.

In an analogous manner, we arrive at a contradiction if we assume that there exist infinitely many k such that

$$\left\|\frac{\partial f(X_k, Y_k)}{\partial Y}\right\|\geqslant\rho''>0.$$

Therefore,

$$F(X(t), Y(t))\underset{t\to\infty}{\longrightarrow}0.$$

Then, for an arbitrary convergent sequence,

$$\{X(t_k), Y(t_k)\}, X(t_k)\underset{t_k\to\infty}{\longrightarrow}X^*; Y(t_k)\underset{t_k\to\infty}{\longrightarrow}Y^*.$$

Here, $[X^*, Y^*]\in M(X_0, Y_0)$. We have

$$\left(\frac{\partial f(X^*, Y^*)}{\partial X}\right)^2+\left(\frac{\partial f(X^*, Y^*)}{\partial Y}\right)^2=2F(X^*, Y^*)=0,$$

that is, the point $[X^*, Y^*]$ satisfies (2.4) and hence is a saddle point.

It remains for us to prove that this saddle point is unique. Suppose that, in addition to $[X^*, Y^*]$, there exists another saddle point $[\overline{X}, \overline{Y}]$ distinct from $[X^*, Y^*]$, that is,

$$\frac{\partial f(\overline{X}, \overline{Y})}{\partial X}=\frac{\partial f(\overline{X}, \overline{Y})}{\partial Y}=0.$$

By virtue of the strong concavity-convexity of $f(X, Y)$, we have

$$f(X, Y^*) < f(X^*, Y^*) < f(X^*, Y) \tag{2.16}$$

for all $X \neq X^*$ and $Y \neq Y^*$ and

$$f(X, \overline{Y}) < f(\overline{X}, \overline{Y}) < f(\overline{X}, \overline{Y}) \tag{2.17}$$

for all $X \neq \overline{X}$ and $Y \neq \overline{Y}$.

Let us consider the three cases:

$$\begin{aligned}
&1)\ X^* \neq \overline{X},\quad Y^* = \overline{Y};\\
&2)\ X^* = \overline{X},\quad Y^* \neq \overline{Y};\\
&3)\ X^* \neq \overline{X},\quad Y^* \neq \overline{Y}.
\end{aligned}$$

Cases 1) and 2) are impossible because, with case 1), we have from (2.16) and (2.17)

$$f(X^*, Y^*) > f(\overline{X}, Y^*) = f(\overline{X}, \overline{Y}) > f(X^*, \overline{Y}) = f(X^*, Y^*),$$

and, with case 2),

$$f(X^*, Y^*) < f(X^*, \overline{Y}) = f(\overline{X}, \overline{Y}) < f(\overline{X}, Y^*) = f(X^*, Y^*).$$

With case 3), we have the incompatible inequalities

$$\begin{aligned}
&f(X^*, Y^*) > f(\overline{X}, Y^*) > f(\overline{X}, \overline{Y}),\\
&f(X^*, Y^*) < f(X^*, \overline{Y}) < f(\overline{X}, \overline{Y}).
\end{aligned}$$

This completes the proof of the theorem.

Systems (2.5) and (2.7) give a "continuous" method of finding saddle points anywhere in the space. On the basis of this "continuous" method, we can construct a number of discrete methods for finding saddle points. Let us look at one of the various possible methods of doing this. Let us take an arbitrary pair $[X_1, Y_1] \in E_n \times E_m$. Let us suppose that $M(X_1, Y_1)$ is bounded. Suppose that we have found X_k and Y_k. Let us look at the rays

$$\begin{aligned}
&X_{k\alpha} = X_k + \alpha G_{Xk},\quad \alpha \in [0, \infty),\\
&Y_{k\beta} = Y_k - \beta G_{Yk},\quad \beta \in [0, \infty),
\end{aligned}$$

where

$$G_{Xk}=\frac{\partial f(X_k, Y_k)}{\partial X} \text{ and } G_{Yk}=\frac{\partial f(X_k, Y_k)}{\partial Y}.$$

It then follows from (2.12) that, for $\frac{dX(\alpha)}{d\alpha}=G_{Xk}$ and $\frac{dY(\beta)}{d\beta}=-G_{Yk}$, we obtain, by considering the vector-valued functions $X(\alpha)$ and $Y(\beta)$ instead of $X(t)$ and $Y(t)$,

$$F(X_{k\alpha}, Y_{k\beta})=F(X_k, Y_k)+\alpha A_k+(\beta-\alpha)B_k-\beta C_k+o_k(\alpha, \beta),$$

where

$$A_k=\left(\frac{\partial f(X_k, Y_k)}{\partial X}, \frac{\partial^2 f(X_k, Y_k)}{\partial X^2}\frac{\partial f(X_k, Y_k)}{\partial Y}\right),$$

$$B_k=\left(\frac{\partial f(X_k, Y_k)}{\partial X}, \frac{\partial^2 f(X_k, Y_k)}{\partial X\partial Y}\frac{\partial f(X_k, Y_k)}{\partial Y}\right),$$

$$C_k=\left(\frac{\partial f(X_k, Y_k)}{\partial Y}, \frac{\partial^2 f(X_k, Y_k)}{\partial Y^2}\frac{\partial f(X_k, Y_k)}{\partial Y}\right),$$

$$\frac{o_k(\alpha, \beta)}{\sqrt{\alpha^2+\beta^2}} \xrightarrow[\substack{\alpha\to+0\\ \beta\to+0}]{} 0,$$

Here, $A_k<0$ if $G_{Xk}\neq 0$ and $C_k>0$ if $G_{Yk}\neq 0$.

If $B_k<0$, we set $\beta=2\alpha$; if $B_k\geqslant 0$, we set $\beta=\frac{1}{2}\alpha$.

Then, if

$$2F(X_k, Y_k)=G_{Xk}^2+G_{Yk}^2>0,$$

we have for sufficiently small α and β

$$F(X_{k\alpha}, Y_{k\beta})<F(X_k, Y_k).$$

Let us find $\alpha_k\in[0, \infty)$ from the condition

$$F(X_{k\alpha_k}, Y_{k\beta(\alpha_k)})=\min_{\alpha\in[0, \infty)} F(X_{k\alpha}, Y_{k\beta(\alpha)})$$

and let us set

$$X_{k+1}=X_{k\alpha_k}, \quad Y_{k+1}=Y_{k\beta(\alpha_k)}.$$

From this point we proceed in an analogous manner. Just as in Theorem 2.2, we can show that $X_k \xrightarrow[k\to\infty]{} X^*$ and $Y_k \xrightarrow[k\to\infty]{} Y^*$ and that $[X^*, Y^*]$ is a saddle point of the function $f(X, Y)$ on $E_n\times E_m$.

Remark 1. Just as in the usual gradient method, we do not have to seek the minimum of $F(X_{k\alpha}, Y_{k\beta(\alpha)})$ on $[0, \infty)$ at each stage but instead we may set

$$X_{k+1} = X_{k\alpha_k} \text{ and } Y_{k+1} = Y_{k\beta(\alpha_k)},$$

where $\alpha_k \in [\varepsilon_0, \varepsilon_1]$, the numbers $\varepsilon_1 > \varepsilon_2 > 0$ being fixed numbers independent of k.

Remark 2. In the method expounded, we can speed up the convergence by setting

$$X_{k+1} = X_{k\alpha_k} \text{ and } Y_{k+1} = Y_{k\beta_k},$$

where α_k and β_k are nonnegative numbers such that

$$F(X_{k\alpha_k}, Y_{k\beta_k}) = \min_{\alpha, \beta \in [0, \infty)} F(X_{k\alpha}, Y_{k\beta}).$$

Here, at each stage, it is necessary to solve the minimization problem with respect to two parameters, which can be laborious.

Remark 3. We can avoid finding the coefficient B_k if this proves too laborious. Then, we need at each stage to test both $\beta = \beta(\alpha) = 2\alpha$ and $\beta = \beta(\alpha) = \frac{1}{2}\alpha$ and then choose whichever of these is appropriate. In this method, there is no need to find the second derivatives.

The search for a saddle point is equivalent to the search for a minimum with respect to X and Y of the function $F(X, Y)$ and we can apply the gradient methods for minimizing $F(X, Y)$ on $E_n \times E_m$. However, to calculate the gradient $F(X, Y)$, it is necessary at each stage to find the second derivatives of the function $f(X, Y)$, which may prove to be laborious.

Remark 4. It is tempting to find X_{k+1} and Y_{k+1} from the conditions $X_{k+1} = X_{k\alpha_k}$ and $Y_{k+1} = Y_{k\beta_k}$, where

$$f(X_{k\alpha_k}, Y_{k\beta_k}) = \min_{\beta \in [0, \infty)} \max_{\alpha \in [0, \infty)} f(X_{k\alpha}, Y_{k\beta}).$$

However, the sequences constructed according to this method may fail to converge to a saddle point.

Remark 5. The method mentioned in Remark 2 can also be applied to the case when $f(X, Y)$ is not a concave-convex function. However, we can arrive at a local minimum of the function $F(X, Y)$.

3°. Let $\Omega_X \subset E_n$ and let $\Omega_Y = E_m$. Suppose that these subsets are strongly convex closed bounded sets. Then, let us consider the functions

$$\psi(X, Y) = \max_{Z \in \Omega_X} \left(\frac{\partial f(X, Y)}{\partial X}, Z - X \right), \tag{2.18}$$

$$\varphi(X, Y) = \min_{Z \in \Omega_Y} \left(\frac{\partial f(X, Y)}{\partial Y}, Z - Y \right) \tag{2.19}$$

for all $[X, Y] \in \Omega_X \times \Omega_Y : \psi(X, Y) \geqslant 0, \ \varphi(X, Y) \leqslant 0.$

Since $\Omega_X \times \Omega_Y$ are strongly convex sets, it follows that, for fixed X and Y, there exist unique points $\theta_1(X, Y)$ in Ω_X and $\theta_2(X, Y)$ in Ω_Y such that

$$\psi(X, Y) = \left(\frac{\partial f(X, Y)}{\partial X}, \theta_1(X, Y) - X \right),$$

$$\varphi(X, Y) = \left(\frac{\partial f(X, Y)}{\partial X}, \theta_2(X, Y) - Y \right).$$

The vector-valued functions $\theta_1(X, Y)$ and $\theta_2(X, Y)$ are continuous on $\Omega_X \times \Omega_Y$.

Here, for simplicity, we shall assume that

$$\frac{\partial f(x, y)}{\partial x} \neq 0, \quad \frac{\partial f(x, y)}{\partial y} \neq 0 \text{ on } \Omega_x \times \Omega_y.$$

Consider the system of differential equations

$$\frac{dX(t)}{dt} \equiv \dot{X}(t) = \theta_1(X(t), Y(t)) - X(t), \tag{2.20}$$

$$X(0) = X_0, \tag{2.21}$$

$$\frac{dY(t)}{dt} \equiv \dot{Y}(t) = \theta_2(X(t), Y(t)) - Y(t), \tag{2.22}$$

$$Y(0) = Y_0. \tag{2.23}$$

If $X_0 \in \Omega_X$ and $Y_0 \in \Omega_Y$, then, for $t \in [0, \infty)$, the solutions $X(t) \equiv X(t, X_0, Y_0)$ and $Y(t) \equiv Y(t, X_0, Y_0)$ of the systems (2.20) and (2.22) [Peano's theorem tells us that a continuous solution exists] belong respectively to the sets Ω_X and Ω_Y. Let $[X(t), Y(t)]$ denote an arbitrary continuous solution of the systems (2.20) and (2.22) with initial conditions (2.21) and (2.23). (Since the right-hand numbers of (2.20) and (2.22) may not satisfy a Lipschitz condition, a unique solution of these systems for arbitrary initial conditions $[X_0\ Y_0]$ in $\Omega_X \times \Omega_Y$ is not ensured.) Then, the functions

$$\psi_1(t) \equiv \psi(X(t),\ Y(t)),$$
$$\varphi_1(t) \equiv \varphi(X(t),\ Y(t))$$

are, as one can show, continuously differentiable on $(0,\ \infty)$ and

$$\frac{d\psi_1(t)}{dt} = \left(\frac{\partial^2 f(t)}{\partial X^2}\dot{X}(t) + \frac{\partial^2 f(t)}{\partial X \partial Y}\dot{Y}(t),\ \theta_1(t) - X(t)\right) - \left(\frac{\partial f(t)}{\partial X},\ \dot{X}(t)\right), \tag{2.24}$$

$$\frac{d\varphi_1(t)}{dt} = \left(\frac{\partial^2 f(t)}{\partial Y \partial X}\dot{X}(t) + \frac{\partial^2 f(t)}{\partial Y^2}\dot{Y}(t),\ \theta_2(t) - Y(t)\right) - \left(\frac{\partial f(t)}{\partial Y},\ \dot{Y}(t)\right), \tag{2.25}$$

where $f(t) = f(X(t),\ Y(t))$ and $\theta_i(t) = \theta_i(X(t),\ Y(t))$ for $i = 1,\ 2$.

THEOREM 2.3. *If $f(X,\ Y)$ is a strongly concave-convex function, then the solutions of the systems (2.20), (2.22) converge, for $[X_0\ Y_0] \in \Omega_X \times \Omega_Y$, to a unique saddle point.*

Proof: Consider the function

$$H(X,\ Y) = \psi(X,\ Y) - \varphi(X,\ Y). \tag{2.26}$$

We note that this function is nonnegative and that it attains its minimum value (which is equal to zero) at a saddle point of the function $f(X,\ Y)$ on $\Omega_X \times \Omega_Y$ (and only at that point). We have, in view of (2.20) and (2.22),

$$h(t) \equiv H(X(t),\ Y(t)) = H(X_0,\ Y_0) + \int_0^t \Big[\Big(\theta_1(\tau) - X(\tau),\ \frac{\partial^2 f(\tau)}{\partial X^2}(\theta_1(\tau) - X(\tau)\Big) +$$
$$+ \left(\theta_1(\tau) - X(\tau),\ \frac{\partial^2 f(\tau)}{\partial X \partial Y}(\theta_2(\tau) - Y(\tau))\right) - \left(\frac{\partial f(\tau)}{\partial X},\ \theta_1(\tau) - X(\tau)\right) -$$
$$- \left(\theta_2(\tau) - Y(\tau),\ \frac{\partial^2 f(\tau)}{\partial Y \partial X}(\theta_1(\tau) - X(\tau))\right) -$$
$$- \left(\theta_2(\tau) - Y(\tau),\ \frac{\partial^2 f(\tau)}{\partial Y^2}(\theta_2(\tau) - Y(\tau))\right) +$$
$$+ \left(\frac{\partial f(\tau)}{\partial Y},\ \theta_2(\tau) - Y(\tau)\right)\Big] d\tau = H(X_0,\ Y_0) +$$
$$+ \int_0^t \left[\left(\dot{X}(\tau),\ \frac{\partial^2 f(\tau)}{\partial X^2}\dot{X}(\tau)\right) - \left(\dot{Y}(\tau),\ \frac{\partial^2 f(\tau)}{\partial Y^2}\dot{Y}(\tau)\right) - \psi(\tau) + \varphi(\tau)\right] d\tau, \tag{2.27}$$

where

$$f(\tau)=f(X(\tau),\ Y(\tau)),\ \theta_i(\tau)=\theta_i(X(\tau),\ Y(\tau))\ (i=1,\ 2),$$
$$\psi(\tau)=\psi(X(\tau),\ Y(\tau)),$$
$$\varphi(\tau)=\varphi(X(\tau),\ Y(\tau)).$$

Let us show that

$$h(t)\underset{t\to\infty}{\longrightarrow}0. \tag{2.28}$$

Let us suppose the opposite, namely, that there exists a sequence $\{t_k\}$ such that

$$\left(t_k\underset{t\to\infty}{\longrightarrow}\infty\right),$$

and

$$h(t_k)\geqslant\rho>0\quad(k=1,\ 2,\ \ldots).$$

For example, suppose that there exist infinitely many k such that

$$\psi(t_k)=\psi(X(t_k),\ Y(t_k))\geqslant\rho'>0.$$

Then,

$$\|\theta_1(X(t_k),\ Y(t_k))-X(t_k)\|\geqslant\rho''>0.$$

All the terms under the integral sign in the right-hand member of (2.27) are nonpositive. By virtue of the properties of the systems (2.20) and (2.22), there exists, for arbitrary $\varepsilon>0$, a $\delta>0$ such that

$$\|X(t)-X(t')\|\leqslant\varepsilon,\quad\text{and}\quad\|Y(t)-Y(t')\|\leqslant\varepsilon$$

whenever t' and t are two nonnegative numbers such that $|t'-t|\leqslant\delta$. Then, there exists a positive number μ such that

$$\|\theta_1(X(t),\ Y(t))-X(t)\|\geqslant\frac{1}{2}\rho''>0$$

for $|t_k-t|\leqslant\mu$.

Keeping (2.9) in mind and reasoning as in the proof of Theorem 2.2, we obtained the result that

$$h(t) \underset{t\to\infty}{\to} -\infty,$$

which is impossible.

In an analogous manner, we reach a contradiction if we assume that there exist infinitely many values of k such that

$$\varphi(t_k) \leqslant -\bar{\rho} < 0.$$

Now, it is easy to show that every convergent sequence $\{X_k, Y_k\}$ converges to a saddle point and that the complete solution $X(t)$, $Y(t)$ approaches a saddle point.

On the basis of the "continuous" method (2.20) and (2.22) of finding a saddle point, we can develop discrete methods for finding saddle points. We present one of these.

Let us choose arbitrarily $X_1 \in \Omega_X$ and $Y_1 \in \Omega_Y$. Suppose that we have found $X_k \in \Omega_X$ and $Y_k \in \Omega_Y$. Suppose that

$$\theta_{1k} = \theta_1(X_k, Y_k);$$

$$\theta_{2k} = \theta_2(X_k, Y_k).$$

If $H(X_k, Y_k) = 0$, then the point $[X_k, Y_k]$ is a saddle point and the process is terminated. On the other hand, if $H(X_k, Y_k) > 0$, let us consider the segment in Ω_X

$$X_{k\alpha} = X_k + \alpha(\theta_{1k} - X_k), \ \alpha \in [0, 1], \ X_{k\alpha} \in \Omega_X$$

and the segment in Ω_Y

$$Y_{k\beta} = Y_k + \beta(\theta_{2k} - Y_k), \ \beta \in [0, 1]; \ Y_{k\beta} \in \Omega_Y.$$

From (2.27), we have

$$h_1(\alpha, \beta) = H(X_{k\alpha}, Y_{k\beta}) = H(X_k, Y_k) + \alpha A_k + \\ + (\beta - \alpha) B_k - \beta C_k + O_k(\alpha, \beta), \tag{2.29}$$

where

$$A_k = \left(\theta_{1k} - X_k, \frac{\partial^2 f(X_k, Y_k)}{\partial X^2} \cdot (\theta_{1k} - X_k)\right),$$
$$B_k = \left(\theta_{2k} - Y_k, \frac{\partial^2 f(X_k, Y_k)}{\partial Y \partial X} (\theta_{1k} - X_k)\right),$$
$$C_k = \left(\theta_{2k} - Y_k, \frac{\partial^2 f(X_k, Y_k)}{\partial Y^2} (\theta_{2k} - Y_k)\right),$$
$$\frac{o_k(\alpha, \beta)}{\sqrt{\alpha^2 + \beta^2}} \xrightarrow[\substack{\alpha \to +0 \\ \beta \to +0}]{} 0$$

uniformly with respect to k.

If $B_k < 0$, we set $\beta = 2\alpha$. In this case, let us consider the function $h_2(\alpha) \equiv h_1(\alpha, 2\alpha)$ and let us find $\alpha_k \in \left[0, \frac{1}{2}\right]$ such that

$$h_2(\alpha_k) = \min_{\alpha \in \left[0, \frac{1}{2}\right]} h_2(\alpha).$$

Let us set

$$X_{k+1} = X_k + \alpha_k (\theta_{1k} - X_k), \tag{2.30}$$
$$Y_{k+1} = Y_k + 2\alpha_k (\theta_{2k} - Y_k). \tag{2.31}$$

On the other hand, if $B_k \geqslant 0$, we set $\beta = \frac{\alpha}{2}$ and consider the function $h_3(\alpha) \equiv h_1\left(\alpha, \frac{1}{2}\alpha\right)$. Let us find $\alpha_k \in [0, 1]$ such that

$$h_3(\alpha_k) = \min_{\alpha \in [0, 1]} h_3(\alpha).$$

Let us set

$$X_{k+1} = X_k + \alpha_k (\theta_{1k} - X_k), \tag{2.32}$$
$$Y_{k+1} = Y_k + \frac{1}{2}\alpha_k (\theta_{2k} - Y_k). \tag{2.33}$$

Obviously, in both cases $X_{k+1} \in \Omega_X$ and $Y_{k+1} \in \Omega_Y$. Also, if $H(X_k, Y_k) > 0$, then $H(X_{k+1}, Y_{k+1}) < H(X_k, Y_k)$. Thus, we construct two sequences $\{X_k\}$ and $\{Y_k\}$.

The sequence $\{H_k\}$, where $H_k = H(X_k, Y_k)$, is monotonically decreasing and hence it converges. Define

$$H^* = \lim_{k \to \infty} H_k.$$

Then,

$$H_k \geqslant H^*.$$

THEOREM 2.4. *The sequences* $\{X_k\}$ *and* $\{Y_k\}$ *constructed above converge to a saddle point of the function* $f(X, Y)$ *on the set* $\Omega_X \times \Omega_Y$.

Proof: Let us show first that

$$H_k \underset{k\to\infty}{\longrightarrow} 0.$$

Let us suppose the opposite. Then, there exists a subsequence $\{H_{k_l}\}$ such that $H_{k_l} \geqslant \rho > 0$ for all $l = 1, 2, \ldots$ where

$$\left(k_l \underset{l\to\infty}{\longrightarrow} \infty\right).$$

Without loss of generality, we may assume that

$$\psi(X_{k_l}, Y_{k_l}) \geqslant \rho' > 0 \tag{2.35}$$

and that

$$B_{k_l} < 0.$$

Consider the segments

$$X_{k_l\alpha} = X_{k_l} + \alpha(\theta_{1k_l} - X_{k_l}), \quad Y_{k_l\alpha} = Y_{k_l} + 2\alpha(\theta_{2k_l} - Y_{k_l}),$$
$$\alpha \in \left[0, \frac{1}{2}\right], \quad X_{k_l\alpha} \in \Omega_X, \quad Y_{k_l\alpha} \in \Omega_Y.$$

From (2.27), we obtain by setting $X(\alpha) = X_{k_l\alpha}$ and $Y(\alpha) = Y_{k_l\alpha}$,

$$\begin{gathered}
\dot{X}(\alpha) = \theta_{1k_l} - X_{k_l}, \quad Y(\alpha) = 2(\theta_{2k_l} - Y_{k_l}), \\
X(0) = X_k, \quad Y(0) = Y_k, \\
h_{k_l}(\alpha) \equiv H(X_{k_l\alpha}, Y_{k_l\alpha}) = H(X_{k_l}, Y_{k_l}) +
\end{gathered}$$
$$\begin{aligned}
&+ \int_0^\alpha \left[\left(\theta_{1k_l} - X_{k_l}, \frac{\partial^2 f(\tau)}{\partial X^2}(\theta_1(\tau) - X(\tau))\right) - \right. \\
&- 2\left(\theta_{2k_l} - Y_{k_l}, \frac{\partial^2 f(\tau)}{\partial Y^2}(\theta_2(\tau) - Y(\tau))\right) + \\
&+ 2\left(\theta_{1k_l} - X_{k_l}, \frac{\partial^2 f(\tau)}{\partial X \partial Y}(\theta_2(\tau) - Y(\tau))\right) - \\
&- \left(\theta_1(\tau) - X(\tau), \frac{\partial^2 f(\tau)}{\partial X \partial Y}(\theta_{2k_l} - Y_{k_l})\right) - \\
&\left. - \left(\frac{\partial f(\tau)}{\partial X}, \theta_{1k_l} - X_{k_l}\right) + 2\left(\frac{\partial f(\tau)}{\partial Y}, \theta_{2k_l} - Y_{k_l}\right)\right] d\tau,
\end{aligned} \tag{2.36}$$

where

$$f(\tau)=f(X_{k_l\tau},\ Y_{k_l\tau}),\ \theta_i(\tau)=\theta_i(X_{k_l\tau},\ Y_{k_l\tau})\ (i=1,\ 2),$$
$$X(\tau)=X_{k_l\tau},\ \ Y(\tau)=Y_{k_l\tau}.$$

By virtue of the boundedness of Ω_X and the continuous differentiability of $f(X,\ Y)$, if follows from (2.35) that

$$\|\theta_{1k_l}-X_{k_l}\|\geqslant\rho''>0.$$

For $\alpha\in(0,\ \alpha_0]$, where α_0 is a positive number independent of k_l, we have from (2.36)

$$\begin{aligned} h_{k_l}(\alpha)\leqslant H(X_{k_l},\ Y_{k_l})&+\\ +\int_0^\alpha\left[\theta_{1k_l}-X_{k_l},\ \frac{\partial^2 f(\tau)}{\partial X^2}(\theta_1(\tau)-X(\tau))\right]d\tau&\leqslant\\ \leqslant H(X_{k_l},\ Y_{k_l})-\frac{1}{2}\alpha m_1\rho''^2=H^*+\varepsilon_{k_l}&-\frac{1}{2}m_1\rho''^2, \end{aligned} \tag{2.37}$$

where $\varepsilon_{k_l}\xrightarrow[l\to\infty]{}0$ and m_1 is a positive number corresponding to the set $\Omega_X\times\Omega_Y$.

Let us fix some $\alpha'\in(0,\ \alpha_0]$ and then define

$$\alpha_0'=\min\left\{\frac{1}{2},\ \alpha_0\right\}.$$

Since

$$\varepsilon_{k_l}\xrightarrow[k_l\to\infty]{}0,$$

it follows that, for $k_l\geqslant K<\infty$, we have

$$h_{k_l}(\alpha)\leqslant H^*-\frac{1}{4}\alpha' m_1\rho''^2<H^*. \tag{2.38}$$

Then,

$$H_{k_l+1}=\min_{\alpha\in\left[0,\frac{1}{2}\right]}h_{k_l}(\alpha)\leqslant h_{k_l}(\alpha')<H^* \tag{2.39}$$

which (2.39) contradicts (2.34).

In a similar manner, we arrive at a contradiction if we assume that $B_{k_l} \geqslant 0$ for all k_l or

$$\varphi(X_{k_l}, Y_{k_l}) \leqslant -\rho < 0$$

for all k_l. Thus, $H_k \xrightarrow[k\to\infty]{} 0$.

It is now clear that all convergent subsequences of the sequence $\{X_k, Y_k\}$ converge to a saddle point. The uniqueness of the saddle point is proven just as in Theorem 2.2.

Remark 1. Remarks 1 - 5 of subsection 2° remain valid, with the obvious modifications, for the case in which the ranges of values of X and Y are bounded sets in the corresponding spaces.

Remark 2. The method expounded corresponds to the "conditional-gradient" method of Sec. 1 of Chapter 3. It is possible to construct methods corresponding to the methods of "projection of the gradient" that were expounded in Sec. 2 of Chapter 3.

Remark 3. In [58, Chapters 6 - 8], we considered the case in which Ω_X and Ω_Y are defined by the inequalities

$$x^i \geqslant 0, \; y^j \geqslant 0 \;\; (i=1, \ldots, n; \; j=1, \ldots, m),$$

where $X=(x^1, \ldots, x^n)$ and $Y=(y^1, \ldots, y^m)$, and a continuous method of finding a saddle point was studied. In this case, instead of the system (2.20) - (2.23), we can consider simpler systems of differential equations (which are studied in detail in the book cited).

Remark 4. The material of this subsection can be carried over without difficulty in the case in which one of the sets Ω_X or Ω_Y is bounded and the other coincides with the entire space or to the case in which Ω_X and Ω_Y are convex and closed though not bounded.

3. Differentiation of a function.

1°. Let $f(X, Y, Z)$ denote a function that is continuous with respect to all three variables, where

$$X \in \Omega_X \subset E_n, \; Y \in \Omega_Y \subset E_m, \; Z \in \Omega_z \subset E_p.$$

On Ω_z, let us consider the function

$$\varphi(Z) = \max_{X \in \Omega_X} \min_{Y \in \Omega_Y} f(X, Y, Z). \tag{3.1}$$

Let us suppose that Ω_X and Ω_Y are closed bounded sets, that Ω_Z is closed, bounded, and convex, and that

$$\left\| \frac{\partial f(X, Y, Z)}{\partial Z} \right\|$$

is bounded on $\Omega_X \times \Omega_Y \times \Omega_Z$. (Of course, we are also assuming that f is continuously differentiable with respect to Z.)

Let us fix some $Z \in \Omega_Z$. Let g denote a nonzero member of E_p such that for a positive number α_0 and a number α in $[0, \alpha_0]$, the point $Z_\alpha = Z + \alpha g$ belongs to Ω_Z. We need to find

$$\varphi'_z(g) = \lim_{\alpha \to +0} \frac{\varphi(Z+\alpha g) - \varphi(Z)}{\alpha}. \tag{3.2}$$

We define two sets $R(Z)$ and $Q(X, Z)$ as follows: $R(Z)$ is the subset of Ω_x consisting of all members X of Ω_X that satisfy the relation

$$\min_{Y \in \Omega_Y} f(X, Y, Z) = \max_{X \in \Omega_X} \min_{Y \in \Omega_Y} f(X, Y, Z);$$

$Q(X, Z)$ is the subset of Ω_Y consisting of all members Y of Ω_Y that satisfy the relation

$$f(X, Y, Z) = \min_{Y \in \Omega_Y} f(X, Y, Z).$$

In this section, we shall assume that the sets $Q(X, Z)$ satisfy the following stringent condition: for given Z, the set $Q(X, Z)$ is a continuous function of X on the set $R(Z)$ [and $Q(X', Z) \to Q(X, Z)$ as $X' \to X$, where X' and X belong to $R(Z)$]. This means that

$$\rho(Q(X', Z), Q(X, Z)) \equiv \sup_{Y \in Q(X', Z)} \inf_{Y \in Q(X, Z)} \|V - Y\| + \\ + \sup_{V \in Q(X, Z)} \inf_{Y \in Q(X', Z)} \|V - Y\| \xrightarrow[X' \to X]{} 0.$$

We note that our condition is satisfied if $Q(X, Z)$ consists, for all $X \in R(Z)$, of a single point.

2°. We define

$$\Phi(X, Z) \equiv \min_{Y \in \Omega_Y} f(X, Y, Z).$$

We note that

$$f(X, Y, Z+\alpha g) = f(X, Y, Z) + \int_0^\alpha \left(\frac{\partial f(X, Y, Z+\theta g)}{\partial Z}, g \right) d\theta. \tag{3.3}$$

Then

$$\varphi_z'(g) = \lim_{\alpha \to +0} \frac{1}{\alpha} \max_{X \in \Omega_X} \min_{Y \in \Omega_Y} \Big[f(X, Y, Z) - \varphi(Z) + $$

$$+ \int_0^\alpha \left(\frac{\partial f(X, Y, Z + \theta g)}{\partial Z}, g \right) d\theta \Big] \equiv \lim_{\alpha \to +0} \frac{1}{\alpha} \max_{X \in \Omega_X} \min_{Y \in \Omega_Y} H(X, Y, \alpha).$$

Since, for every $X' \in R(Z)$,

$$\max_{X \in \Omega_X} \min_{Y \in \Omega_X} H(X, Y, \alpha) \geqslant \min_{Y \in \Omega_Y} H(X', Y, \alpha),$$

it follows that, for $X' \in R(Z)$,

$$\lim_{\alpha \to +0} \frac{1}{\alpha} \max_{X \in \Omega_X} \min_{Y \in \Omega_Y} H(X, Y, \alpha) \geqslant \lim_{\alpha \to +0} \frac{1}{\alpha} \min_{Y \in \Omega_Y} H(X', Y, \alpha). \quad (3.4)$$

It follows from (II.3.31) that $\varphi(Z) = \Phi(X, Z)$ for $X \in R(Z)$ and

$$\lim_{\alpha \to +0} \frac{1}{\alpha} \min_{Y \in \Omega_Y} H(X', Y, \alpha) = \min_{Y \in Q(X', Z)} \left(\frac{\partial f(X', Y, Z)}{\partial Z}, g \right).$$

Therefore,

$$\frac{\partial \varphi(Z)}{\partial g} \geqslant \max_{X \in R(Z)} \min_{Y \in Q(X, Z)} \left(\frac{\partial f(\dot{X}, Y, Z)}{\partial Z}, g \right). \quad (3.5)$$

On the other hand,

$$\min_{Y \in \Omega_Y} H(X, Y, \alpha) \leqslant \min_{Y \in Q(X, Z)} H(X, Y, \alpha) =$$

$$= \min_{Y \in Q(X, Z)} \left[f(X, Y, Z) - \varphi(Z) + \int_0^\alpha \left(\frac{\partial f(X, Y, Z + \theta g)}{\partial Z}, g \right) d\theta \right] =$$

$$= \Phi(X, Z) - \varphi(Z) + \min_{Y \in Q(X, Z)} \int_0^\alpha \left(\frac{\partial f(X, Y, Z + \theta g)}{\partial Z}, g \right) d\theta, \quad (3.6)$$

because $f(X, Y, Z) = \Phi(X, Z)$ for $y \in Q(X, Z)$.

From (3.6) we obtain

$$\max_{X \in \Omega_X} \min_{Y \in \Omega_Y} H(X, Y, Z) \leqslant \max_{X \in \Omega_X} \Big[\Phi(X, Z) - \varphi(Z) +$$

$$+ \min_{Y \in Q(X, Z)} \int_0^\alpha \left(\frac{\partial f(X, Y, Z + \theta g)}{\partial Z}, g \right) d\theta \Big] \equiv \max_{X \in \Omega_X} W(X, Z, \alpha). \quad (3.7)$$

For arbitrary fixed $\varepsilon > 0$, let us consider the sets $X_{1\varepsilon}$ consisting of all X that belong to Ω_X and satisfy the inequality

$$|\Phi(X, Z) - \varphi(Z)| \leqslant \varepsilon, \tag{3.8}$$

and its complement $X_{2\varepsilon} = \Omega_X \setminus X_{1\varepsilon}$, that is, the set $X_{2\varepsilon}$ consisting of all X in Ω_X that satisfy the inequality

$$|\Phi(X, Z) - \varphi(Z)| > \varepsilon.$$

In the present case, inequality (3.8) is equivalent to the inequalities

$$-\varepsilon \leqslant \Phi(X, Z) - \varphi(Z) \leqslant 0. \tag{3.9}$$

Since

$$\max_{X \in \Omega_X} \left[\Phi(X, Z) - \varphi(Z) + \min_{Y \in Q(X, Z)} \int_0^\alpha \left(\frac{\partial f(X, Y, Z + \theta g)}{\partial Z}, g \right) d\theta \right] =$$
$$= \max \left\{ \sup_{X \in X_{1\varepsilon}} W(X, Z, \alpha), \sup_{X \in X_{2\varepsilon}} W(X, Z, \alpha) \right\},$$

one can easily see that for sufficiently small $\alpha > 0$,

$$\max_{X \in \Omega_X} W(X, Z, \alpha) = \sup_{X \in X_{1\varepsilon}} W(X, Z, \alpha) \equiv$$
$$\equiv \sup_{X \in X_{1\varepsilon}} \left[\Phi(X, Z) - \varphi(Z) + \min_{Y \in Q(X, Z)} \int_0^\alpha \left(\frac{\partial f(X, Y, Z + \theta g)}{\partial Z}, g \right) d\theta \right].$$

Since

$$\sup_{X \in X_{1\varepsilon}} [A(X) + B(x)] \leqslant \sup_{X \in X_{1\varepsilon}} A(X) + \sup_{X \in X_{1\varepsilon}} B(X),$$

and since, by virtue of (3.9),

$$\Phi(X, Z) - \varphi(Z) \leqslant 0,$$

it follows that, for these $\alpha > 0$,

$$\max_{X \in \Omega_X} W(X, Z, \alpha) = \sup_{X \in X_{1\varepsilon}} W(X, Z, \alpha) \leqslant$$
$$\leqslant \sup_{X \in X_{1\varepsilon}} \min_{Y \in Q(X, Z)} \int_0^\alpha \left(\frac{\partial f(X, Y, Z + \theta g)}{\partial Z}, g \right) d\theta.$$

From this inequality and inequality (3.7), we have

$$\varphi_Z'(g) = \lim_{\alpha\to+0} \frac{1}{\alpha} \max_{X\in\Omega_X} \min_{Y\in\Omega_Y} H(X, Y, \alpha) \leqslant$$

$$\leqslant \lim_{\alpha\to+0} \frac{1}{\alpha} \max_{X\in\Omega_X} W(X, Z, \alpha) \leqslant \sup_{X\in X_{1\varepsilon}} \max_{Y\in Q(X, Z)} \left(\frac{\partial f(X, Y, Z)}{\partial Z}, g\right).$$

Let us now let ε approach 0. Keeping the continuity of the set $Q(X, Z)$ with respect to X on $R(Z)$ in mind, we obtain

$$\varphi_Z'(g) \leqslant \max_{X\in R(Z)} \min_{Y\in Q(X, Z)} \left(\frac{\partial f(X, Y, Z)}{\partial Z}, g\right). \tag{3.10}$$

From (3.10) and (3.5), we finally obtain

$$\varphi_Z'(g) = \max_{X\in R(Z)} \min_{Y\in Q(X, Z)} \left(\frac{\partial f(X, Y, Z)}{\partial Z}, g\right). \tag{3.11}$$

Remark 1. Formula (3.11) remains valid when $\Omega_X = E_n$ and $\Omega_Y = E_m$ except that we need to impose the following restrictions:

1) There exists a positive number ρ such that the set

$$\bigcup_{Z\in S_\rho} R(Z)$$

is bounded, where $S_\rho \subset E_p$ is the sphere of positive radius ρ with center at the point Z, and for no $Z \in S_\rho$ does there exist a sequence $\{X_j\}$ such that $\|X_j\| \to \infty$ and

$$\Phi(X_j, Z) \to \sup_{X\in E_n} \Phi(X, Z).$$

2) For arbitrary $X \in E_n$ and $Z \in S_\rho$, the sets $Q(X, Z)$ and

$$\bigcup_{\substack{X\in E_n \\ Z\in S_\rho}} Q(X, Z)$$

are bounded, and for no combination of $X \in E_n$ and $Z \in S_\rho$ does there exist a sequence $\{Y_i\}$ such that $\|Y_i\| \to \infty$; $Y_i \in E_m$, and

$$f(X, Y_i, Z) \underset{i\to\infty}{\longrightarrow} \inf_{Y\in E_m} f(X, Y, Z).$$

3) $\left\|\frac{\partial f(X, Y, Z)}{\partial Z}\right\|$ is bounded on $E_n \times E_m \times S_\rho$.

If we seek an extremum of the function $\varphi(Z)$ on E_m, it is necessary to require in addition that the above three conditions hold for $\rho = \infty$ (that is, we need to take E_p instead of S_ρ).

Remark 2. Let $f(X_1, \ldots, X_s, Z$ denote a given function and suppose that $X_i \in \Omega_i \subset E_{n_i}$ (for $i = 1, \ldots, s$) and $Z \in \Omega_Z \subset E_p$.

Let us suppose that the function f is continuous with respect to all $s+1$ variables and continuously differentiable with respect to Z. Suppose that

$$\varphi(Z) = \min_{X_1 \in \Omega_1} \max_{X_2 \in \Omega_2} \min_{X_3 \in \Omega_3} \ldots f(X_1, \ldots, X_s, Z).$$

Suppose that $R_1(Z)$ is a subset of Ω_1 such that, for $X_1' \in R_1(Z)$,

$$\varphi_1(X_1', Z) = \min_{X_1 \in \Omega_1} \varphi_1(X_1, Z),$$

where

$$\varphi_1(X_1, Z) = \max_{X_2 \in \Omega_2} \min_{X_3 \in \Omega_3} \ldots f(X_1, \ldots, X_s, Z).$$

Suppose, also, that the set $R_2(X_1, Z)$ is a subset of Ω_2 such that, for $X_2' \in R_2(X_1, Z)$,

$$\varphi_2(X_1, X_2', Z) = \max_{X_2 \in \Omega_2} \varphi_2(X_1, X_2, Z),$$

where

$$\varphi_2(X_1, X_2, Z) = \min_{X_3 \in \Omega_3} \ldots f(X_1, \ldots, X_s, Z).$$

Analogously, we define the sets

$$R_3(X_1, X_2, Z) \subset \Omega_3;\ R_4(X_1, X_2, X_3, Z) \subset \Omega_4.$$

Then, if the set $R_2(X_1, Z)$ is continuous with respect to X_1 on the set $R_1(Z)$ and if the set $R_3(X_1, X_2, Z)$ is continuous with respect to X_1 and X_2 on the set $\{X_1, X_2\}$, where $X_1 \in R_1(Z)$ and $X_2 \in R_2(X_1, Z)$, etc., it follows by induction that, just as above,

$$\varphi_Z'(g) = \min_{X_1 \in R_1(Z)} \max_{X_2 \in R_2(X_1, Z)} \min_{X_3 \in R_3(X_1, X_2, Z)} \ldots \left(\frac{\partial f(X_1, \ldots, X_s, Z)}{\partial Z}, g\right). \tag{3.12}$$

4. Differentiation of a function (continued).

1°. Let $f(X, Y, Z)$ denote a continuous function defined on $E_n \times E_m \times E_p$. Let us fix $Z \in E_p$. Let $R(Z)$ denote the subset of E_n consisting of all X in E_n such that

$$\min_{Y \in E_m} f(X, Y, Z) = \max_{X \in \Omega_X} \min_{Y \in \Omega_Y} f(X, Y, Z) \equiv \varphi(Z),$$

Let $Q(X, Z)$ denote the subset of Ω consisting of all Y in Ω_Y such that

$$f(X, Y, Z) = \min_{Y \in \Omega_Y} f(X, Y, Z) \equiv \Phi(X, Z).$$

We make the following assumptions:

1) For $Z \in E_p$ there does not exist a sequence of points $\{X_j\}$ such that $\|X_j\| \to \infty$ and

$$\Phi(X_j, Z) \to \sup_{X \in E_n} \Phi(X, Z).$$

2) For arbitrary $X \in E_n$, the sets $Q(X, Z)$ and

$$\bigcup_{\substack{X \in E_n \\ Z \in S_\varepsilon}} Q(X, Z)$$

are bounded. Also, for no combination of $X \in E_n$ and $Z \in S_\varepsilon$ there exists a sequence $\{Y_i\}$ such that $\| Y_i \| \to \infty$, where each $Y_i \in E_m$ and

$$f(X, Y_i, Z) \underset{i \to \infty}{\longrightarrow} \inf f(X, Y, Z).$$

Here, S_ε is the sphere of radius ε in E_p with center at the point Z.

3) The function $f(X, Y, Z)$ is twice continuously differentiable with respect to X and Z on $E_n \times E_m \times S_\varepsilon$ and it is strictly concave with respect to X for arbitrary fixed $Y \in E_m$ and $Z \in S_\varepsilon$. Also, the derivatives

$$\left\| \frac{\partial f(X, Y, Z)}{\partial X} \right\|, \left\| \frac{\partial f(X, Y, Z)}{\partial Y} \right\|, \left\| \frac{\partial^2 f(X, Y, Z)}{\partial X^2} \right\|,$$
$$\left\| \frac{\partial^2 f(X, Y, Z)}{\partial Z^2} \right\|, \left\| \frac{\partial^2 f(X, Y, Z)}{\partial X \partial Z} \right\|$$

are bounded on $E_n \times E_m \times S_\varepsilon$. Let g denote a nonzero member of E_p. There exists

$$\varphi'_Z(g) = \lim_{\alpha \to +0} \frac{1}{\alpha} [\varphi(Z + \alpha g) - \varphi(Z)]. \tag{4.1}$$

Remark 1. The problem posed above differs from the problem studied in the preceding section in that here we do not assume that the set $Q(X, Z)$ is continuous with respect to X on $R(Z)$.

2°. Let us evaluate $\varphi'_Z(g)$. We have

$$\varphi'_Z(g) = \lim_{\alpha \to +0} \frac{1}{\alpha} \left[\max_{X \in E_n} \min_{Y \in E_m} (f(X, Y, Z + \alpha g) - \varphi(Z)) \right].$$

For arbitrary fixed $X \in R(Z)$ and $V \in E_n$,

$$\max_{X \in E_n} \min_{Y \in E_m} [f(X, Y, Z+\alpha g) - \varphi(Z)] \geqslant$$
$$\geqslant \min_{Y \in E_m} [f(X+\alpha V, Y, Z+\alpha g) - \varphi(Z)] =$$
$$= \min_{Y \in E_m} \left\{ f(X, Y, Z) - \varphi(Z) + \int_0^\alpha \left[\left(\frac{\partial f(X+\theta V, Y, Z+\theta g)}{\partial Z}, g \right) + \right.\right.$$
$$\left.\left. + \left(\frac{\partial f(X+\theta V, Y, Z+\theta g)}{\partial X}, V \right) \right] d\theta \right\}.$$

Therefore, for arbitrary $X \in R(Z)$ and $V \in E_n$,

$$\varphi'_Z(g) \geqslant \lim_{\alpha \to +0} \frac{1}{\alpha} \min_{Y \in E_m} \{ f(X, Y, Z) - \varphi(Z) +$$
$$+ \int_0^\alpha \left[\left(\frac{\partial f(X+\theta V, Y, Z+\theta g)}{\partial Z}, g \right) + \left(\frac{\partial f(X+\theta V, Y, Z+\theta g)}{\partial X}, V \right) \right] d\theta. \quad (4.2)$$

Since

$$\varphi(Z) = \Phi(X, Z) = \min_{Y \in E_m} f(X, Y, Z),$$

for $X \in R(Z)$, it follows from the definition of a derivative with respect to direction that the right-hand member of (4.2) is the derivative with respect to directions $g' \in E_{n+p}$ (where $g' = [V, g]$) of the function

$$\varphi_1(W) = \min_{Y \in E_m} f_1(Y, W),$$

where $f_1(Y, W) = f(X, Y, Z)$ and $W = [X, Z] \in E_{n+p}$. In fact,

$$\varphi'_W(g') = \lim_{\alpha \to +0} \frac{1}{\alpha} [\varphi_1(W+\alpha g') - \varphi_1(W)] =$$
$$= \lim_{\alpha \to +0} \frac{1}{\alpha} \left[\min_{Y \in E_m} f(X+\alpha V, Y, Z+\alpha g) - \varphi_1(W) \right].$$

Since $X \in R(Z)$, we have

$$\varphi_1(W) = \min_Y f(X, Y, Z) = \Phi(X, Z) = \varphi(Z),$$
$$f(X+\alpha V, Y, Z+\alpha g) = \int_0^\alpha \left[\left(\frac{\partial f(X+\theta V, Y, Z+\theta g)}{\partial Z}, g \right) + \right.$$
$$\left. + \left(\frac{\partial f(X+\theta V, Y, Z+\theta g)}{\partial X}, Y \right) \right] d\theta,$$

It is obvious that the right-hand member of inequality (4.2) is equal to

$$\varphi'_W(g') = \min_{Y \in Q(X, Z)} \left[\left(\frac{\partial f(X, Y, Z)}{\partial Z}, g \right) + \left(\frac{\partial f(X, Y, Z)}{\partial X}, V \right) \right].$$

Therefore, remembering that X and V are arbitrary except that X is a member of $R(Z)$ and V is a member of E_n, we have

$$\varphi'_Z(g) \geqslant \max_{X \in R(Z)} \max_{V \in E_n} \min_{Y \in Q(X, Z)} \left[\left(\frac{\partial f(X, Y, Z)}{\partial Z}, g \right) + \right.$$
$$\left. + \left(\frac{\partial f(X, Y, Z)}{\partial X}, V \right) \right]. \quad (4.3)$$

Let us now prove the converse assertion, namely, that

$$\varphi'_Z(g) = \lim_{\alpha \to +0} \frac{1}{\alpha} \max_{X \in E_n} \min_{Y \in E_m} [f(X, Y, Z + \alpha g) - \varphi(Z)]. \quad (4.4)$$

It is easy to show that there exists a set $R^*(Z, g) \subset R(Z)$ such that

$$R(Z + \alpha g) \xrightarrow[\alpha \to +0]{} R^*(Z, g),$$

where the convergence of the sets is to be understood in the sense indicated in subsection 1° of Sec. 3. It then follows that for every point $\overline{X} \in R^*(Z, g)$ and every α in an interval of the form $(0, \alpha_0]$, where $\alpha_0 > 0$, there exists an $X(\alpha) \in R(Z + \alpha g)$, so that, for an arbitrary sequence $\{\alpha_i\}$, where each $\alpha_i \in (0, \alpha_0]$, that approaches 0, we have $X(\alpha) \to \overline{X}$. The point $X(\alpha)$ can be represented in the form

$$X(\alpha) = \overline{X} + \alpha V(\alpha), \text{ where } \|\alpha V(\alpha)\| \xrightarrow[\alpha \to +0]{} 0.$$

Then, for $\overline{X} \in R^*(Z, g)$,

$$\max_{X \in E_n} \min_{Y \in E_m} [f(X, Y, Z + \alpha g) - \varphi(Z)] =$$
$$= \min_{Y \in E_m} [f(X(\alpha), Y, Z + \alpha g) - \varphi(Z)] =$$
$$= \min_{Y \in E_m} [f(\overline{X} + \alpha V(\alpha), Y, Z + \alpha g) - \varphi(Z)] =$$
$$= \min_{Y \in E_m} \left\{ f(\overline{X}, Y, Z) - \varphi(Z) + \alpha \left[\left(\frac{\partial f(\overline{X}, Y, Z)}{\partial X}, V(\alpha) \right) + \right. \right.$$
$$\left. + \left(\frac{\partial f(\overline{X}, Y, Z)}{\partial Z}, g \right) \right] +$$

$$+\frac{1}{2}\alpha^2\left[\left(\frac{\partial^2 f(\overline{X}+\theta(\alpha)V(\alpha),\ Y,\ Z+\theta(\alpha)g)}{\partial X^2}V(\alpha),\ V(\alpha)\right)+\right.$$
$$+2\left(\frac{\partial^2 f(\overline{X}+\theta(\alpha)V(\alpha),\ Y,\ Z+\theta(\alpha)g)}{\partial X\partial Z}V(\alpha),\ Z\right)+$$
$$\left.\left.+\left(\frac{\partial^2 f(X+\theta(\alpha)V(\alpha),\ Y,\ Z+\theta(\alpha)g)}{\partial Z^2}g,\ g\right)\right]\right\}\equiv$$
$$\equiv\min_{Y\in E_m}\left\{f(\overline{X},\ Y,\ Z)-\varphi(Z)+\alpha A+\frac{1}{2}\alpha^2 B\right\}\leqslant$$
$$\leqslant\min_{Y\in Q(\overline{X},\ Z)}\left\{f(\overline{X},\ Y,\ Z)-\varphi(Z)+\alpha A+\frac{1}{2}\alpha^3 B\right\}=$$
$$=\min_{Y\in Q(\overline{X},\ Z)}\left[\alpha A+\frac{1}{2}\alpha^2 B\right], \tag{4.5}$$

because $Y\in Q(\overline{X},\ Z)$ $\overline{X}\in R^*(Z,\ g)\subset R(Z)$, and $f(\overline{X},\ Y,\ Z)=\varphi(Z)$. In (4.5), $\theta(\alpha)\in[0,\ \alpha]$.

Since

$$\min_{Y\in Q(\overline{X},\ Z)}\left[\alpha A+\frac{1}{2}\alpha^2 B\right]\leqslant\min_{Y\in Q(X,\ Z)}\alpha A+\max_{Y\in Q(X,\ Z)}\frac{1}{2}\alpha^2 B,$$

we have *a fortiori*

$$\min_{Y\in Q(\overline{X},\ Z)}\left[\alpha A+\frac{1}{2}\alpha^2 B\right]\leqslant$$
$$\leqslant\max_{V\in E_n}\min_{Y\in Q(\overline{X},\ Z)}\alpha\left[\left(\frac{\partial f(\overline{X},\ Y,\ Z)}{\partial X},\ V\right)+\left(\frac{\partial f(\overline{X},\ Y,\ Z)}{\partial Z},\ g\right)+\right.$$
$$+\max_{V\in E_n}\max_{Y\in Q(\overline{X},\ Z)}\frac{1}{2}\alpha^2\left[\left(\frac{\partial^2 f(\overline{X}+\theta(\alpha)V(\alpha),\ Y,\ Z+\theta(\alpha)g)}{\partial X^2}V,\ V\right)+\right.$$
$$+2\left(\frac{\partial^2 f(\overline{X}+\theta(\alpha)V(\alpha),\ Y,\ Z+\theta(\alpha)g)}{\partial X\partial Z}V,\ g\right)+$$
$$\left.+\left(\frac{\partial^2 f(\overline{X}+\theta(\alpha)V(\alpha),\ Y,\ Z+\theta(\alpha)g)}{\partial Z^2}g,\ g\right)\right]\equiv$$
$$\equiv\max_{V\in E_n}\min_{Y\in Q(\overline{X},\ Z)}\alpha\overline{A}+\max_{V\in E_n}\max_{Y\in Q(\overline{X},\ Z)}\frac{1}{2}\alpha^2\overline{B}. \tag{4.6}$$

Since

$$\overline{X}+\theta(\alpha)V(\alpha)\xrightarrow[\alpha\to+0]{}\overline{X}\text{ and } Z+\theta(\alpha)g\xrightarrow[\alpha\to+0]{}0$$

and since $\frac{\partial^2 f}{\partial X^2}$ is a negative-definite matrix (by virtue of the assumption of strict concavity of $f(X,\ Y,\ Z)$ with respect to X for arbitrary Y and Z), it follows that, for sufficiently small $\alpha\in[0,\ \alpha_1]$, where $\alpha_1 > 0$, there exists a number K in $(0, \infty)$, independent of $\alpha \in [0, \alpha_0]$ and $Y \in Q(X, Z)$, such that

$$\left(\frac{\partial^2 f(\overline{X} + \theta(\alpha) V(\alpha), Y, Z + \theta(\alpha) g)}{\partial X^2} V, V\right) \leqslant K \| V \|^2.$$

Then,

$$\left| \max_{V \in E_n} \max_{Y \in Q(\overline{X}, Z)} \overline{B} \right| \leqslant L < \infty. \tag{4.7}$$

Therefore, from (4.4) – (4.7), we have

$$\varphi'_Z(g) \leqslant \max_{V \in E_n} \min_{Y \in Q(\overline{X}, Z)} \left[\left(\frac{\partial f(\overline{X}, Y, Z)}{\partial X}, V\right) + \left(\frac{\partial f(\overline{X}, Y, Z)}{\partial Z}, g\right)\right]. \tag{4.8}$$

Since (4.8) holds for all $\overline{X} \in R^*(Z, g)$ and $R^*(Z, g) \subset R(Z)$, we have

$$\varphi'_Z(g) \leqslant \max_{X \in R(Z)} \max_{V \in E_n} \min_{Y \in Q(X, Z)} \left[\left(\frac{\partial f(X, Y, Z)}{\partial X}, V\right) + \left(\frac{\partial f(X, Y, Z)}{\partial Z}, g\right)\right]. \tag{4.9}$$

From (4.3) and (4.9), we have finally,

$$\varphi'_Z(g) = \max_{X \in R(Z)} \max_{V \in E_n} \min_{Y \in Q(X, Z)} \left[\left(\frac{\partial f(X, Y, Z)}{\partial X}, V\right) + \left(\frac{\partial f(X, Y, Z)}{\partial Z}, g\right)\right]. \tag{4.10}$$

Remark 1. If

$$\frac{\partial^2 f}{\partial X \partial Z} \equiv 0 \equiv \psi(X, Z) \equiv 0,$$

for some $Y \in Q(\overline{X}, Z)$, it is sufficient to require that the function $f(X, Y, Z)$ be concave (rather than strictly concave).

Remark 2. Formula (4.10) remains valid, as one can easily see, when $Y \in \Omega_Y \subset E_m$, where Ω_Y is a closed bounded subset of E_m.

Remark 3. Suppose that $X \in \Omega_X \subset E_n$ is a closed bounded subset of E_n. Repeating the reasoning used in the derivation of (4.10) (and keeping in mind the fact that Ω_X is bounded), we obtain the following expression for the derivative with respect to direction in the present case:

$$\varphi'_Z(g) = \max_{X \in R(Z)} \max_{V \in E_n} \min_{Y \in Q(X, Z)} \left[\left(\frac{\partial f(X, Y, Z)}{\partial X}, V \right) + \right.$$
$$\left. + \left(\frac{\partial f(X, Y, Z)}{\partial Z}, g \right) \right]. \tag{4.11}$$

where $M_X(\Omega_X)$ is the cone of admissible directions in the broad sense of the word constructed at the point X from the set Ω_X (see subsection 4° of Sec. 1 of Chapter 2).

Remark 4. If $f(X, Y, Z)$ is defined on $\Omega_X \times \Omega_Y \times \Omega_Z$ and if the sets Ω_X, Ω_Y, and Ω_Z are bounded, it is sufficient to require that $f(X, Y, Z)$ be twice continuously differentiable with respect to X and Z on $\Omega_X \times \Omega_Y \times \Omega_Z$.

Remark 5. If

$$\varphi(Z) = \min_{X \in E_n} \max_{Y \in \Omega_Y} f(X, Y, Z)$$

where Ω_Y is a closed subset of the space E_m, then (4.10) takes the form

$$\varphi_Z(g) = \min_{X \in E(Z)} \min_{V \in E_n} \max_{Y \in Q(X, Z)} \left[\left(\frac{\partial f(X, Y, Z)}{\partial X}, V \right) + \right.$$
$$\left. + \left(\frac{\partial f(X, Y, Z)}{\partial Z}, g \right) \right], \tag{4.12}$$

where Y is a member of $R(Z)$ if

$$\max_{Y \in E_m} f(X, Y, Z) = \min_{V \in E_n} \max_{Y \in E_m} f(X, Y, Z),$$

and Y is a member of $Q(X, Z)$ if

$$f(X, Y, Z) = \max_{W \in \Omega_Y} f(X, W, Z).$$

Remarks 1 – 4 remain valid in the present case (with the corresponding modifications due to the fact that we have min max instead of max min).

The condition of strict concavity with respect to X that we imposed on the function $f(X, Y, Z)$ is replaced with the condition of strict convexity with respect to X of the function $f(X, Y, Z)$.

3°. Example. Consider

$$\varphi(\lambda) = \min_{X \in E_1} \max \{(X - \lambda)^2, X^2\}, \quad \lambda \in E_1.$$

Define

$$\Phi(X, \lambda) = \max\{(X-\lambda)^2, X^2\}.$$

The function $\Phi(X, \lambda)$ can be reduced to the standard form

$$\Phi(X, \lambda) = \max_{Y \in \Omega} f(X, Y, \lambda)$$

in the following way: we set $f(X, Y, \lambda) = (Y-1)(X-\lambda)^2 + (2-Y)X^2$ and $\Omega = \{1, 2\}$. Obviously,

$$f(X, 1, \lambda) = X^2, \ f(X, 2, \lambda) = (X-\lambda)^2.$$

In a similar way, the function

$$\Phi(X, Z) = \max_{i=\{1, 2, \ldots, N)} f_i(X, Z)$$

can be reduced to the form

$$\Phi(X, Z) = \max_{Y \in \Omega} f(X, Y, Z),$$

by setting

$$\Omega = \{1, 2, \ldots, N\};$$

and

$$f(X, Y, Z) = \sum_{i=1}^{N} \omega_i(Y) f_i(X),$$

where

$$\omega_i(Y) = \frac{1}{a_i} \prod_{\substack{i \in \{1, \ldots, N\} \\ j \neq i}} (j-i),$$

and

$$a_i = \prod_{\substack{j \in \{1, \ldots, N\} \\ j \neq i}} (i-j) = (-1)^{N-i}(i-1)!(N-i)!$$

In the present case, it is easy to find $\varphi(\lambda)$:

$$\Phi(X, \lambda)=\begin{cases} X^2 & \text{for } X^2 \geqslant (X-\lambda)^2, \\ (X-\lambda)^2 & \text{for } X^2 \leqslant (X-\lambda)^2. \end{cases}$$

We note that $\Phi(X, \lambda)$ is continuous with respect to X and λ.

For $\lambda > 0$ (see Fig. 8),

$$\Phi(X, \lambda)=\begin{cases} X^2 & \text{for } X \geqslant \frac{1}{2}\lambda, \\ (X-\lambda)^2 & \text{for } X \leqslant \frac{1}{2}\lambda. \end{cases}$$

For $\lambda < 0$,

$$\Phi(X, \lambda)=\begin{cases} X^2 & \text{for } X \leqslant \frac{1}{2}\lambda, \\ (X-\lambda)^2 & \text{for } X \geqslant \frac{1}{2}\lambda. \end{cases}$$

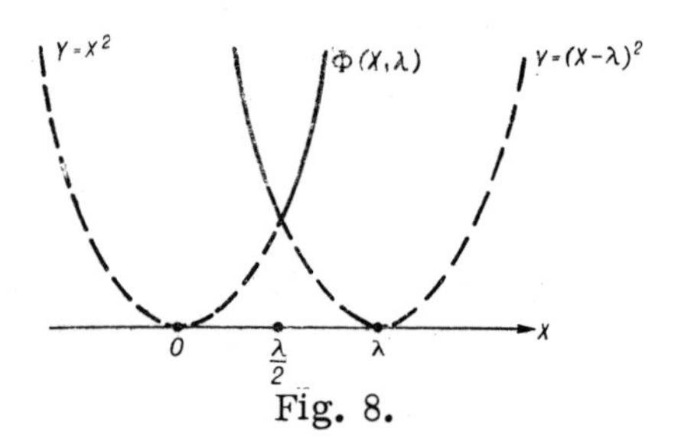

Fig. 8.

For arbitrary λ,

$$\varphi(\lambda)=\frac{1}{4}\lambda^2=\Phi\left(\frac{1}{2}\lambda, \lambda\right),$$

Therefore,

$$\frac{d\varphi(\lambda)}{d\lambda}=\frac{1}{2}\lambda.$$

On the other hand, in accordance with (4.12), we have

$$\dot{\varphi}_+(\lambda)=\varphi'_\lambda(g),$$

where $g=1$. In the present case, $R(\lambda)=\frac{\lambda}{2}$ and $Q(X, Z)=\{1, 2\}$. Therefore, from (4.12)

$$\begin{aligned}\frac{\partial\varphi(\lambda)}{\partial g}&=\min_{V\in E_1}\max\{\lambda V, -\lambda V+\lambda\}=\\&=-\lambda\min_{V\in E_1}\max\{V, 1-V\}.\end{aligned}\tag{4.14}$$

Since

$$\max\{V, 1-V\}=\begin{cases}V, & V\geqslant 1-V,\\ 1-V, & V\leqslant 1-V,\end{cases}$$

that is, since

$$\max\{V, 1-V\}=\begin{cases}V, & V\geqslant\frac{1}{2},\\ 1-V, & V\leqslant\frac{1}{2}.\end{cases}$$

it is obvious that

$$\min_{V\in E_1}\max\{V, 1-V\}=\frac{1}{2}.$$

Thus, from (4.14), we have $\frac{\partial\varphi(\lambda)}{\partial g}=\frac{\lambda}{2}$. Analogously, we have

$$\dot{\varphi}_-(\lambda)=-\varphi_\lambda'(g),$$

where $g=-1$. Repeating for $g=-1$ the line of reasoning given above, we obtain

$$\frac{\partial\varphi(X)}{\partial g}=-\frac{\lambda}{2}$$

so that

$$\dot{\varphi}_-(\lambda)=\dot{\varphi}(\lambda)=\frac{1}{2}\lambda,$$

that is, the function $\varphi(\lambda)$ is differentiable at the point λ and

$$\frac{\partial \varphi(\lambda)}{d\lambda} = \frac{1}{2}\lambda. \tag{4.5}$$

Thus, the value obtained by direct calculation (formula [4.14]) for the derivative $\frac{\partial \varphi(\lambda)}{\partial \lambda}$ coincides with the value (4.15) obtained from formula (4.12).

REFERENCES

1. Amvrosenko, V. V., "Uskoreniye skhodimosti metoda Brauna resheniya matrichnykh igr" (Speeding up of convergence of Brown's method of solving matrix games), *Ekonomika i matematicheskiye metody*, 1, 4, 1965.
2. Akilov, G. P., and A. M. Rubinov, "Metod posledovatel'nykh priblizheniy dlya razyskaniya mnogochlena nailuchshego priblizheniya" (A method of successive approximations for finding the polynomial of best approximation), *Doklady Akad. nauk SSSR*, 157, 3, 1964.
3. Bellman, R., I. Glicksberg, and O. Gross, *Some Aspects of the Mathematical Theory of Control Processes*, The Rand Corporation, Rand Report R-313.
4. Bourbaki, N., Fonctions d'une variable réelle.
5. Vajda, S., *An Introduction to Linear Programming and the Theory of Games*, London, Methuen, and New York, Wiley, 1960.
6. Vainberg, M. M., *Variational Methods for the Study of Nonlinear Operations*, Holden-Day, 1963 (translation of *Variatsionnyye metody issledovaniya nelineynykh operatorov)*.
7. Vertgeym, B. A., and G. Sh. Rubinshteyn, "K opredeleniyu kvazivypuklykh funktsiy" (On the determination of quasiconvex functions) in the collection *Matematicheskoye programmirovaniye*, Moscow, izd. "Nauka," 1966.
8. Volkonskiy, V. A., "Optimal'noye planirovaniye v usloviyakh bol'shoy razmernosti" (Optimal planning under conditions of high dimensionality), *Ekonomika i matematicheskiye metody*, 1, 2, 1965.
9. Gamkrelidze, R. V., "O skol'zyashchikh optimal'nykh rezhimakh" (On sliding optimal regimes), *Doklady Akad. nauk SSSR*, 143, 6, 1962.

10. Gol'shyteyn, Ye. G., "Dvoystvennyye zadachi vypuklogo programmirovaniya" (Dual problems in convex programming), *Ekonomika i matematicheskiye metody*, 1, 3, 1965.
11. Gol'shteyn, Ye. G., "Metody blochnogo programmirovaniya" (Methods of block programming), *Ekonomika i matematicheskiye metody*, 2, 1, 1966.
12. Danskin, D. M., "Iterativnyy metod resheniya nepreryvnykh igr" (Iterative method of solving continuous games) in collection *Beskonechnyye antagonisticheskiye igry*, Fitzmatgiz, 1963.
13. Dem'yanov, V. F., "Postroyeniye programmnogo upravleniya v lineynoy sisteme, optimal'nogo v integral'nom smysle" (The construction of a program control in a linear system that is optimal in the integral sense), *Prikladnaya matematika i mekhanika*, 27, 3, 1963.
14. Dem'yanov, V. F., "K postroyeniyu optimal'noy programmy v lineynyky sistemakh" (On the construction of an optimal program in linear systems), *Avtomatika i telemekhanika*, 25, 1, 1964.
15. Dem'yanov, V. F., "K minimizatsii funktsiy na ogranichennykh mnozhestvakh" (On the minimization of functions defined on bounded sets), *Kibernetika*, 1, 6, 1965.
16. Dem'yanov, V. F., "K nakhozhdeniyu optimal'nykh upravleniy v nekotorykh zadachakh avtomaticheskogo regulirovaniya" (On finding optimal controls in certain automatic control problems), *Vestnik*, LGU (Leningrad State University), No. 13, 1965.
17. Dem'yanov, V. F., "K resheniyu nekotorykh nelineynykh zadach optimal'nogo upravleniya" (On the solution of nonlinear optimal-control problems), *Zhurnal vych. matematiki i matematicheskoy fiziki*, 6, 2, 1966.
18. Dem'yanov, V. F., "K minimizatsii maksimal'nogo ukloneniya" (On the minimization of maximum deviation), *Vestnik*, LGU (Leningrad State University), No. 7, 1966.
19. Dem'yanov, V. F., "Pryamoy metod nakhozhdeniya optimal'nykh upravleniy v nelineynoy sisteme avtomaticheskogo regulirovaniya" (A direct method for finding optimal controls in a nonlinear automatic control system) in the collection *Prikladnyye zadachi tekhnicheskoy kibernetiki*, Moscow, izd. "Sov.Radio," 1966.
20. Dem'yanov, V. F., "K resheniyu nekotorykh minimaksynkh zadach" (On the solution of certain minimax problems), I, *Kibernetika*, 2, 6, 1966.

21. Dem'yanov, V. F., "Optimizatsiya nelineynykh sistem upravleniya pri ogranicheniyakh na fazovyye koordinaty" (Optimization of nonlinear control systems when there are restrictions on the phase coordinates), *Vestnik* LGU (Leningrad State University), No. 13, 1966.
22. Dem'yanov, V. F. and A. M. Rubinov, "Minimizatsiya gladkogo vypuklogo funktsionala na vypuklom mnozhestve" (Minimization of a smooth convex functional defined on a convex set), *Vestnik*, LGU (Leningrad State University), No. 19, 1964.
23. Dem'yanov, V. F., and A. M. Rubinov, "K zadache o minimizatsii gladkogo funktsionala pri vypuklykh ogranicheniyakh" (Problem of minimizing a smooth functional under convex restrictions), *Doklady Akad. nauk SSSR*, 160, 1, 1965.
24. Dem'yanov, V. F., and A. M. Rubinov, "O neobkhodimykh usloviyakh minimuma" (On necessary conditions for a minimum), *Ekonomika i matematicheskiye metody*, 2, 3, 1966.
25. Dubovitskiy, A. Ya., and A. A. Milyutin, "Zadachi na ekstremum pri nalichiii ogranicheniya" (Extremum problems when there is a constraint), *Zhurnal vych. matematiki i matematicheskoy fiziki*, 5, 3, 1965.
26. Ziyaudinova, D. A., and A. M. Rubinov, "Minimizatsiya sublineynykh funktsionalov na vypuklom kompakte v metrizuyemom lokal'novypuklom prostranstve" (Minimization of sublinear functionals defined on a convex compact subset of a metrizable locally convex space) in the collection *Optimal'noye planirovaniye*, No. 7. Novosibirsk, Sibiriskoye otdeleniye izd. "Nauka," 1967.
27. Zoutendijk, G., Methods of Feasible Directions, Elsevier, 1960.
28. Zukhovitskiy, S. I., P. A. Polyak, and M. Ye. Primak, "Chislennyy metod dlya resheniya zadachi vypuklogo programmirovaniya v gil'bertovom prostranstve" (Numerical method for solving a problem of convex programming in a Hilbert space), *Doklady Akad. nauk SSSR*, 163, 2, 1962.
29. Eaton, J. H., "An iterative solution to time-optimal control," *J. Math. Anal. and Appl.*, 5, 2, 1962.
30. Kazarinov, Yu. F., "Optimal'nyye upravleniya v sistemakh s zapazdyvayushchim argumentom (Optimal controls in systems with lagging argument), in the collection *Prikladnyye zadachi tekhnicheskoy kibernetiki*, Moscow, izd. "Sov. Radio," 1966.
31. Kantorovich, L. V., "Ob odnom effektivnom methode resheniya nekotorykh klassov ekstremal'nykh problem" (On an efficient

method of solving certain classes of extremal problems), *Doklady Akad. nauk SSSR*, 28, 1, 1940.

32. Kantorovich, L. V., and K. P. Akilov, *Functional Analysis in Normed Spaces*, New York, Pergamon, 1964 (translation of *Funktsional'nyy analiz v normirovannykh prostranstvakh)*.
33. Karlin, S., *Mathematical Methods and Theory in Game Theory Programming, and Economics*, Reading, Massachusetts, Addison Wesley, 1959.
34. Kirillova, F. M., "The application of functional analysis methods to some control problems," *J. SIAM Control*, 5, 5, 1967.
35. Kirin, N. Ye., "Ob odnom chislennom metode v teorii lineynykh bystrodeystviyakh (On a numerical method in the problem of linear high-speed operations), in the collection *Metody vychisleniy*, No. 2, Leningrad, 1963.
36. Krasnosel'skiy, M. A., *Topological Methods in the Theory of Nonlinear Integral Equations*, New York, Pergamon, 1964 (translation of *Topologicheskiye metody v teorii nelineynykh integral'nykh uravneniy)*.
37. Krein, M. G., "L-problema v abstraktnom lineynom normirovannom prostranstve" (The L-problem in an abstract normed linear space) in the book by N. Akhiyezer and M. Krein. *O nekotorykh voprosakh teorii momentov* (On certain questions in the theory of moments), Khar'kov, 1938.
38. Krylov, I. A., and F. L. Chernous'ko, "O metode posledovatel'nykh priblizheniy dlya resheniya zadach optimal'nogo upravleniya" (On a method of successive approximations for solving optimal-control problems), *Zhurnal vych. matematiki i matematicheskoy fiziki*, 2, 6, 1962.
39. Levitin, Ye. S., B. T. Polyak, "Methody minimizatsii pri nalichii ogranicheniy" (Methods of minimization when there are constraints), *Zhurnal vych. matematiki i matematicheskoy fiziki*, 6, 5, 1966.
40. Lyusternik, L. A., and V. I. Sobolev, *Elements of functional analysis*, Ungar, 1961 (translation of *Elementy funktsional'nogo analiza)*.
41. Moiseyev, N. N. "Chislennyye metody teorii optimal'nykh upravleniy, ispol'zuyushchiye variatsii v prostranstve sostoyaniy" (Numerical methods in optimal-control theory that use variations in the state space), *Kibernetika*, No. 3, 1966.
42. Neustadt, L. W., "Synthesizing time-optimal systems," *J. Math. Anal. and Appl.*, 1, 3, 1960.

43. Neustadt, L. W., "An abstract variational theory with applications to a broad class of optimization problems, 1," *J. SIAM Control*, 4, 3, 1966.
44. Polyak, B. T., "Teoremy sushchestovavniya i skhodimost' minimiziruyushchikh posledovatel'nostey dlya zadach na ekstremum pri nalichii ogranicheniy" (Existence theorem and the convergence of minimizing sequences for extremum problems with constraints), *Doklady Akad. nauk SSSR*, 166, 2, 1966.
45. Pontryagin, L. S., V. G. Boltyanskiy, R. V. Gamkrelidze, and Ye. F. Mishchenko, *Mathematical Theory of Optimal Processes*, New York, Interscience, 1962 and New York, Pergamon, 1964 (Translations of *Matematicheskaya teoriya optimal'nykh protsessov*).
46. Pshenichnyy, B. N., "Chislennyy metod rascheta optimal'nogo po bystrodeystviyu upravleniya dlya lineynykh sistem" (A numerical method for computing the optimal high-speed control for linear systems), *Zhurnal vych. matematiki i matematicheskoy fiziki*, 4, 1, 1964.
47. Pshenichnyy, B. N., "Dvoystrennyy metod v ekstremal'nykh zadachakh" (A dual method in extremal problems), I, II, *Kibernetika*, 1, 3-4, 1965.
48. Pshenichnyy, B. N., "Vypukloye programmirovaniye v normirovannykh prostranstvakh" (Convex programming in normed spaces), *Kibernetika*, 1, 5, 1965.
49. Rastrigin, L. A., *Sluchaynyy poisk v zadachakh optimizatsii mnogoparametricheskikh sistem* (Random search in problems of optimization of many-parameter systems), Riga, izd. "Znaniye," 1965.
50. Robinson, D., "Iterativnyy metod resheniya igr" (An iterative method of solving games) in the collectin *Matrichnyye Igry* (Matrix games), Fizmatgiz, 1961.
51. Rubinov, A. M., "Minimizatsiya normy na kompakte" (Minimization of a norm on a compact set), *Vestnik*, LGU (Leningrad State University), No. 1, 1965.
52. Rubinov, A. M., "O svyazyakh mezhdu matematicheskim programmirovaniyem i nekotorymi nelineynymi zadachami" (On the relationships between mathematical programming and certain nonlinear problems), in the collection *Tez. dokl. Vsesoyuz, mezhvuz. konf. po primeneniyu funktsional'nogo analiza k resheniye nelineynykh zadach* (Synopses of addresses of the All-Union Inter-university Conference on the application of

functional analysis methods to the solution of nonlinear problems), Baku, 1965.

53. Rubinov, A. M., "Neobkhodimyye usloviya ekstremuma i ikh primeneniye k issledovaniyu nekotorykh uravneniy" (Necessary conditions for an extremum and their application to the study of certain equations), *Doklady Akad. nauk SSSR*, 169, 3, 1966.
54. Rubinov, A. M., "O nekotorykh obobshcheniyakh methoda naiskoreyshego spuska" (On some generalizations of the method of fastest descent), in the collection *Matematicheskoye programmirovaniye*, izd. "Nauk," 1966.
55. Rubinov, A. M., "O nekotorykh svoystvakh sublineynykh funktsionalov" (On certain properties of sublinear functionals), in the collection *Optimal'noye planirovaniye*, No. 9, Sibirskoye otdeleniye izd. "Nauk," 1967.
56. Rubinshteyn, G. Sh., "Dvoystvennyye ekstremal'nyye zadachi" (Dual extremal problems), *Doklady Akad. nauk SSSR*, 152, 1963.
57. Frank, M., and P. Wolfe, "An algorithm for quadratic programming," *Naval Research Logistic Quarterly*, 3, 1956.
58. Arrow, K. J., *et al.*, Studies in Linear and Nonlinear Programming, Stanford University Press, 1958.

INDEX